INFLAMMATORY BOWEL DISEASE: CROSSROADS OF MICROBES, EPITHELIUM AND IMMUNE SYSTEMS

The Novartis Foundation is an international scientific and educational charity (UK Registered Charity No. 313574). Known until September 1997 as the Ciba Foundation, it was established in 1947 by the CIBA company of Basle, which merged with Sandoz in 1996, to form Novartis. The Foundation operates independently in London under English trust law. It was formally opened on 22 June 1949.

The Foundation promotes the study and general knowledge of science and in particular encourages international co-operation in scientific research. To this end, it organizes internationally acclaimed meetings (typically eight symposia and allied open meetings and 15–20 discussion meetings each year) and publishes eight books per year featuring the presented papers and discussions from the symposia. Although primarily an operational rather than a grant-making foundation, it awards bursaries to young scientists to attend the symposia and afterwards work with one of the other participants.

The Foundation's headquarters at 41 Portland Place, London W1B 1BN, provide library facilities, open to graduates in science and allied disciplines. Media relations are fostered by regular press conferences and by articles prepared by the Foundation's Science Writer in Residence. The Foundation offers accommodation and meeting facilities to visiting scientists and their societies.

Information on all Foundation activities can be found at http://www.novartisfound.org.uk

Novartis Foundation Symposium 263

INFLAMMATORY BOWEL DISEASE: CROSSROADS OF MICROBES, EPITHELIUM AND IMMUNE SYSTEMS

2004

John Wiley & Sons, Ltd

Published in 2004 by John Wiley & Sons Ltd,
The Atrium, Southern Gate,
Chichester PO19 8SQ, UK

National 01243 779777
International (+44) 1243 779777
e-mail (for orders and customer service enquiries): cs-books@wiley.co.uk
Visit our Home Page on http://www.wileyeurope.com
or http://www.wiley.com

Other Wiley Editorial Offices

John Wiley & Sons Inc., 111 River Street, Hoboken, NJ 07030, USA

Jossey-Bass, 989 Market Street, San Francisco, CA 94103-1741, USA

Wiley-VCH Verlag GmbH, Boschstr. 12, D-69469 Weinheim, Germany

John Wiley & Sons Australia Ltd, 33 Park Road, Milton, Queensland 4064, Australia

John Wiley & Sons (Asia) Pte Ltd, 2 Clementi Loop #02-01, Jin Xing Distripark, Singapore 129809

John Wiley & Sons Canada Ltd, 22 Worcester Road, Etobicoke, Ontario, Canada M9W 1L1

Wiley also publishes its books in a variety of electronic formats. Some content that appears in print may not be available in electronic books.

Novartis Foundation Symposium 263
ix+228 pages, 16 figures, 12 tables

Library of Congress Cataloging-in-Publication Data

Inflammatory bowel disease: crossroads of microbes, epithelium, and immune systems.
p. ; cm. – (Novartis Foundation symposium ; 263)
"Symposium on Inflammatory bowel disease – crossroads of microbes, epithelium, and immune systems, held at the Novartis Foundation, London, 25–27 November 2003.
Editors, Derek Chadwick (organizer) and Jamie Goode"–Contents p.
Includes bibliographical references and index.
ISBN 0-470-09046-4 (alk. paper)
1. Inflammatory bowel diseases – Congresses. I. Chadwick, Derek. II. Goode, Jamie. III. Series.
[DNLM: 1. Inflammatory Bowel Diseases – Congresses. 2. Genetic Predisposition to Disease – Congresses. 3. Intestinal Mucosa – immunology – Congresses. WI 420 I4234 2004]
RC862.I53I525 2004
616.3′44 – dc22
2004057748

British Library Cataloguing in Publication Data

A catalogue record for this book is available from the British Library

ISBN 0 470 09046 4

Typeset in 10½ on 12½ pt Garamond by Dobbie Typesetting Limited, Tavistock, Devon.
Printed and bound in Great Britain by T. J. International Ltd, Padstow, Cornwall.
This book is printed on acid-free paper responsibly manufactured from sustainable forestry, in which at least two trees are planted for each one used for paper production.

Contents

Symposium on Inflammatory bowel disease—crossroads of microbes, epithelium and immune systems, held at the Novartis Foundation, London, 25–27 November 2003

Editors: Derek Chadwick (Organizer) and Jamie Goode

This symposium is based on a proposal made by Derek Jewell and Bill Roediger

Participants

T. Ahmad Gastroenterology Unit, University of Oxford, Gibson Laboratories, Radcliffe Infirmary, Woodstock Road, Oxford OX2 6HE, UK

I. Bjarnason Department of Medicine, Guy's, Kings, St Thomas' Medical School, Bessemer Road, London SE5 9PJ, UK

J. H. Cummings Division of Pathology and Neuroscience, Ninewells Hospital and Medical School, Dundee DD1 9SY, UK

R. N. Fedorak Division of Gastroenterology, University of Alberta, Suite 205, College Plaza, 8215-112 Street, Edmonton, AB T6G 2C8, Canada

E. Furrie MRC Microbiology and Gut Biology Group, University of Dundee Medical School, Dundee DD1 9SY, UK

S. Ghosh Department of Gastroenterology, Imperial College London, Hammersmith Hospital, Ducane Road, London W12 0NN, UK

P. R. Gibson Department of Gastroenterology, Box Hill Hospital, Box Hill, Victoria 3128, Australia

A. Hart (*Novartis Foundation Bursar*) Physiology Unit, St Mark's Hospital, Northwick Park, Watford Road, Harrow HA1 2UJ, UK

D. P. Jewell (*Chair*) Gastroenterology Unit, Radcliffe Infirmary, Gibson Building, 2nd Floor, Woodstock Road, Oxford OX2 6HE, UK

M. A. Kamm St Mark's Hospital, Northwick Park, Watford Road, Harrow HA1 3UJ, UK

D. Kelleher Department of Clinical Medicine, Trinity College, Trinity Health Sciences Building, St James Hospital, Dublin 8, Ireland

Y. R. Mahida Institute of Infection, Immunity and Inflammation and Division of Gastroenterology, Faculty of Medicine and Health Sciences, University Hospital, Queen's Medical Centre, Nottingham NG7 2UH, UK

J. B. Meddings University of Calgary, Faculty of Medicine, Division of Gastroenterology, 3330 Hospital Drive Northwest, Calgary, Alberta, T2N 4N1, Canada

H. Moore GI Diseases, Department of Pharmacology, Novartis Research Institute, Brunner Str 59, A-1230 Wien, Austria

C. A. Parkos Department of Pathology, Emory University, Whitehead Biomedical Building, Room 105B Michael Street, Atlanta, GA 30322, USA

P. Pavli Department of Gastroenterology, The Canberra Hospital, Woden, ACT 2606, Australia

A. S. Peña Department of Gastroenterology, Laboratory of Immunogenetics, University Medical Center, PO Box 7057, 1007 MB Amsterdam, The Netherlands

F. Powrie Sir William Dunn School of Pathology, University of Oxford, South Parks Road, Oxford OX1 3RE, UK

J. M. Rhodes Gastroenterology Research Group, Department of Medicine, University of Liverpool, Liverpool L69 7ZA, UK

J. D. Rioux Inflammatory Disease Research, Broad Institute of MIT and Harvard, Brigham and Women's Hospital, Harvard Medical School, Cambridge, MA 02139-1561, USA

W. E. W. Roediger Department of Surgery, University of Adelaide and Colorectal Unit, The Queen Elizabeth Hospital, Adelaide, SA 5011, Australia

R. B. Sartor Department of Medicine/GI, Room 7309 Biomolecular Research Building, University of North Carolina, CB #7032, Chapel Hill, NC 27599-7032, USA

J. Schölmerich Department of Internal Medicine I, University of Regensburg, 93042 Regensburg, Germany

E. F. Stange Abteilung fur Innere Medizin I, Robert-Bosch-Krankenhaus, Auerbachstr 110, 70376 Stuttgart, Germany

Holm H. Uhlig Sir William Dunn School of Pathology, University of Oxford, South Parks Road, Oxford OX1 3RE, UK

S. J. van Deventer Department of Experimental Internal Medicine, Academic Medical Centre, Meibergdreef 9, 1105 AZ Amsterdam, The Netherlands

Adam Wilkins BioEssays, 10/11 Tredgold Lane, Napier Street, Cambridge CB1 1HN, UK

N. A. Wright Bart's and the London Queen Mary's School of Medicine and Dentistry, Turner Street, London E1 2AD, UK

Chair's introduction

Derek Jewell

Gastroenterology Unit, Radcliffe Infirmary, Gibson Building, 2nd Floor, Woodstock Road, Oxford OX2 6HE, UK

The Novartis Foundation is remarkable. It was formed in 1949 as the Ciba Foundation and one of the people instrumental in setting it up was Lord Beveridge, one of the architects of post-war social policy. During the inaugural meeting he said, 'This place is not a laboratory for mixing compounds, but we do mean to make it a laboratory for mixing scientists.' This has been the philosophy of the Foundation over the last 50 years. The style of the symposia has remained largely unchanged, as small meetings with about 30 people and lots of time for discussion. I hope that the discussion over the next two and a half days will be as free and frank as possible.

It is virtually 100 years since a very famous debate took place in the Royal Society of Medicine. In 1905 there was an acrimonious discussion about the aetiology of ulcerative colitis (UC). Samuel Wilks had recognized that all bloody diarrhoeas were not dysentery in 1859. By 1905 there were three main groups of people with fixed views about what was going on. There was clearly a group who didn't believe that there was such a thing as UC, and it was all a dysenteric illness due to infection. It was about the end of the 19th century that preservatives and additives were being used in the food industry. Not surprisingly there was a group of people who thought colitis was all due to these food additives. Then there was a third group, who maintained it was all in the mind. Here we are 100 years later: have we moved anywhere? Yes, principally through the application of genetics.

In the 1960s and 1970s, when large clinics of UC and Crohn's disease (CD) patients started to develop, it was recognized that these diseases could run in families. But no one knew how to study complex genetic disorders that did not follow Mendelian lines until 1992 when the microsatellite technology was described by Marc Lathrop. In the last decade we have not only known where to look in the genome, but also we have found a gene. Having got a gene, how do we proceed from here? We don't know how it works, and we don't know how it mediates susceptibility to CD, and so the big issue is, once we find other genes, to find out how they work and how the mutations in the genes actually mediate susceptibility to presumably what is in the gut lumen. This brings into play the whole topic of this symposium. We are exploring the interplay between the bacteria in the lumen, the epithelial barrier and the mucosal immune system, in addition to the genetic susceptibility underlying all of this. This is the purpose of this meeting. Hopefully we will all go away having thought of things from a different perspective, with ideas for further research.

Progress towards identifying inflammatory bowel disease susceptibility genes

John D. Rioux

Inflammatory Disease Research, Broad Institute of MIT and Harvard, Brigham and Women's Hospital, Harvard Medical School, Cambridge, MA 02139-1561, USA

Abstract. Genomewide searches have identified about a dozen genomic regions potentially containing inflammatory bowel disease (IBD) susceptibility loci. Although some were observed in more than one genome scan or in replication studies, no one locus has been identified in all studies. This apparent lack of consistency is explained by the modest effect of most genetic loci in complex traits and by the modest power and the sampling variance of individual family-based studies. Despite these challenges, the field of IBD genetics found convincing evidence of linkage and, more recently, of genetic sequence variants that are strongly associated with disease. The first example was that of the *IBD1* locus on chromosome 16, a linkage region confirmed by a large multicentre study, for which the *CARD15* gene was identified—using independent association strategies—as being a gene contributing to Crohn's disease susceptibility. Despite these advances, a number of questions still remain: (1) What is the best approach to identify the remaining IBD susceptibility loci? (2) How do we prove causality for any given gene or genetic variant? (3) How can we best use this information to obtain an understanding of the biological mechanisms underlying disease susceptibility and to identify useful markers of disease progression and response to therapy?

2004 Inflammatory bowel disease—crossroads of microbes, epithelium and immune systems. Wiley, Chichester (Novartis Foundation Symposium 263) p 3–16

Crohn's disease (CD) and ulcerative colitis (UC) are idiopathic, chronic inflammatory disorders of the gastrointestinal tract that arise in genetically susceptible individuals. To understand the challenge of identifying the genetic variation that confers disease susceptibility, it is important to know the general characteristics of the human genome and of disease susceptibility genes. Specifically, it is known that the size of the human genome is ~3 billion base pairs, of which ~10 million are commonly variable in the human population. Inflammatory bowel diseases (IBD), like other complex human traits, are believed to result from the interplay between a number of disease genes and environmental factors. The specific genetic variants of these disease genes are expected to individually confer modest disease risk and thus are found in healthy individuals (albeit at a lower frequency) in addition to IBD patients. In an effort to localize these genetic variants, much effort has been spent during the last few years performing genomewide searches for regions of linkage. To date, a dozen genomewide searches have been performed on more than 2000 affected relative pairs with IBD from ~1200 families. Six genomic regions showing the most consistent evidence of linkage have been designated *IBD1–IBD6* (MIM#s 266600, 601458, 604519, 606675, 606348 and 606674 respectively). It should be noted that no single study identified all loci and, conversely, no single locus was identified in all studies. This apparent lack of consistency is easily explained by the modest power of individual family collections (~100–200 families) and the significant effect of sampling variance on the distribution of disease alleles between studies. This is best illustrated by the first true replication of an IBD susceptibility locus (the *IBD1* locus on chromosome 16, LOD score = 5.79) that resulted from pooling genotype data from over 613 families collected by 12 centres distributed throughout North America, Europe, and Australia. It was striking that the evidence for linkage in this study came from all 12 centres, including those that had not previously found convincing linkage to this locus (Cavanaugh 2001).

In general, the regions of linkage identified by these genomewide scans are 20–30 centimorgans (cM) (over 20 million base pairs); therefore, it is important to be able to refine these localizations efficiently. In order to demonstrate a comprehensive strategy for achieving this aim, I will review our identification of the causal genetic variation in the linked region on 5q31, known as *IBD5*, which contributes susceptibility to IBD. *IBD5* was first identified in a genomewide linkage scan of 183 affected relative pairs with IBD from 158 Canadian families (Rioux et al 2000). In addition to statistical evidence of genomewide significance for the *IBD5* locus, this region also contains the cytokine gene cluster—a cluster of genes of biological relevance. This finding led us to focus substantial effort on this region and embark on a hierarchical linkage disequilibrium (LD) mapping study of the *IBD5* locus (Rioux et al 2001). Specifically, we typed 56 microsatellite markers across the 18 cM linked region in 256 father–mother–affected child trios and identified two markers with significant association to CD ($P < 0.0005$), as tested with the transmission disequilibrium test (TDT). Analysis of the sequence in the 1 Mb region surrounding these two markers identified additional microsatellite markers. Typing of these markers confirmed the association signal ($P < 0.0005$). Multi-marker analyses identified a six-marker haplotype that bounded the associated region to ~500 kb. Mutation screening of all of the genes in the region revealed no likely causal sequence variants. We therefore undertook a comprehensive sequence analysis of the entire critical region. We re-sequenced eight individuals at every base in the 285 kb core region (i.e. where association was greatest) and selected regions (about every other 500 bp

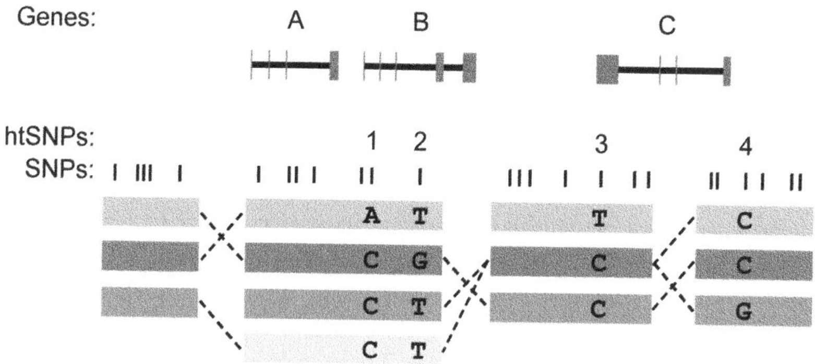

FIG. 1. General features of the haplotype structure of the human genome. Single nucleotide polymorphisms (SNPs; represented by vertical black tick marks) are randomly spread throughout the genome. Regions with little or no evidence of historical recombination are depicted as solid blocks. The alleles of SNPs within a given block form a limited number of patterns or haplotypes (represented by different shades). Some of the SNPs will have an allele that is unique for a given haplotype and are known as haplotype tagging SNPs (htSNPs). Dashed lines between blocks represent the recombination events that lead to the mosaic pattern of haplotypes found in actual chromosomes. Currently there is no known relationship between gene and haplotype block locations—some blocks may contain >1 gene and some genes are located across >1 block. The challenges of determining causality can also be illustrated in this figure: if the A and T alleles at htSNPs #1 and #3 respectively, were equally significantly associated with a disease phenotype, the lack of recombination between these alleles would make it difficult to establish which of these SNPs and which of the three genes is causal.

segment) that totalled 185 kb on either side of the core region. A total of 651 single nucleotide polymorphisms (SNPs) were identified and most have since been typed in trios with CD.

Rather than finding a single causal variant, however, we identified 11 SNPs that were significantly associated ($P < 0.00005$) with CD (we have since found about a dozen more). All of the SNPs had essentially identical information and could individually account for the linkage and association observed at this locus. Examination of the patterns of alleles (or haplotypes) in the control chromosomes (untransmitted parental chromosomes) revealed that there were long segments of LD separated by intervals bearing evidence of multiple recombination events (Daly et al 2001) (see Fig. 1). These segments, or haplotype blocks, contained a striking lack of diversity (2–4 haplotypes accounted for 90–98% of all chromosomes) considering the large number of possible combinations of these alleles. This haplotype structure therefore suggested that—from a biological as well as a statistical perspective—the question of association in this region should simply be to test each haplotype in each block for evidence of association. Performing this test revealed that the most common haplotype across six adjacent blocks had significant excess transmission (all other haplotypes were undertransmitted). With essentially perfect haplotype information, simulations demonstrated that the disease locus has a 90% probability of being within the 250 kb region over which the transmitted to untransmitted ratio is at least 2.25:1. This risk haplotype is very common: its estimated frequency in the general population is ~37% but it is carried by over 75% of all CD patients studied (Rioux et al 2001).

In genetic analyses of complex human traits it is essential to replicate, in an independent cohort, any statistically significant finding of association for confirmation of that finding (Lander & Kruglyak 1995). We therefore performed studies in two independent cohorts and found significant evidence of replication (Rioux et al 2001, Giallourakis et al 2003). Three additional studies were reported by other groups, and all found significant evidence of replication (Armuzzi et al 2003, Mirza et al 2003, Negoro et al 2003). Taken together, these studies include data from over 2000 CD cases and matched controls (family- and population-based), thus providing overwhelming evidence of replication ($P < 5 \times 10^{-10}$). These studies also demonstrate that the relative risk for this associated haplotype (OR 1.31; 95% CI 1.20–1.42) is likely to be more modest than the estimate from the original study, at least partially explained by the phenomenon known as the 'Winner's Risk' (Hirschhorn et al 2002). It should also be noted that *IBD5* represents the second confirmed example of genetic variants influencing susceptibility to CD. The first CD susceptibility gene, known as *CARD15* (formerly *NOD2*) was discovered by Hugot and colleagues using an LD mapping approach (Hugot et al 2001) and by Ogura and colleagues using a candidate gene approach (Ogura et al 2001). Association to this gene, located within the *IBD1*-linked region on chromosome 16q, was subsequently confirmed in multiple studies (Hampe et al 2001, Abreu et al 2002, Ahmad et al 2002, Vavassori et al 2002, Vermeire et al 2002a, Cavanaugh et al 2003). It is worth underlining the fact that not all groups that found significant association to *IBD5* or *CARD15* had actually detected evidence of linkage in the same population; providing empirical support for the notion that association studies are more powerful than linkage studies when attempting to identify susceptibility genes for common disease (Risch & Merikangas 1996).

At this point it is worth examining the more general implications of this haplotype structure.

- If one or more genes are contained in a single block, and that block has X haplotypes, even if the haplotype block contains many dozens of SNPs and other genetic variants, the genes really only have X common alleles or the genetic equivalent of isoforms.
- Common haplotypes capture most of the genetic variation across sizable regions, including the variants that have yet to be discovered within a block.

- It is usually possible to identify an SNP that has an allele that is uniquely found on a single haplotype (all other haplotypes bearing the other allele). These are called haplotype-tagging SNPs or htSNPs (see Fig. 1). Selecting htSNPs (Johnson et al 2001) or combinations of SNPs that uniquely tag each haplotype within a block for an association study allows you to type a small number of SNPs, and yet test all of the variation within the block.
- This structure clarifies the role of physical distance in associations between a marker and a disease allele. An htSNP that uniquely tags the haplotype bearing the disease allele will be more informative than a closer marker that does not uniquely tag that disease haplotype (Fig. 1). In addition, this structure also clarifies our observation of multiple *IBD5* SNPs that are equally associated with CD: all were htSNPs tagging the *IBD5* risk haplotype. Any one, or a combination, of those htSNPs could be conferring the increased risk to developing CD, but since they were all in near perfect LD, it is not possible at this stage to determine which is causal. It should also be noted that since many of the genes in this region have already been shown to be differentially expressed between colon samples from CD and healthy individuals, additional biological evidence would be necessary to establish causality. Furthermore, one cannot rule out the possibility that the risk conferred by this locus acts via coordinated expression of multiple genes.

Taken together, observations regarding the *IBD5* haplotype as well as observations regarding other genomic regions (Daly et al 2001, Jeffreys et al 2001, Johnson et al 2001, Rioux et al 2001) suggest that a haplotype-based strategy is a more comprehensive approach to association studies than randomly testing an arbitrary number of SNPs (Goldstein 2001). These studies, however, only examine a limited number of genomic regions using SNPs identified by re-sequencing. Therefore, it is important that a subsequent large-scale study by Gabriel and colleagues confirmed this structure for an additional 60 regions using publicly available SNP databases (Gabriel et al 2002). Their study shows that haplotype blocks are a general feature of the genomic landscape, with much more than 80% of the entire genome falling into very tightly linked blocks. These blocks are consistently large in size across the genome—in fact, half of all randomly ascertained 20 kb genomic segments show no evidence of historical recombination and consequently display very low haplotype diversity in a population of European descent. Gabriel and colleagues also examined the haplotype structure and diversity for these same regions in a few Asian-derived and African-derived populations and demonstrated that the Asian-derived samples resemble the Caucasian-derived data and that the African-derived samples have a greater diversity of mutations and recombination events (as evidenced by greater numbers of haplotypes per block and smaller block sizes, respectively), consistent with the ages of these populations (Gabriel et al 2002). Importantly, Gabriel and colleagues also demonstrated that most of this variation is shared across all populations; only a small fraction of this diversity is specific to a given population.

Given the interest in using the haplotype structure of the human genome to identify disease genes, it is important to determine whether known causal variants are detectable using common SNPs identified in control (i.e. non-patient) DNA samples. One such test case is the *CARD15* gene, as it was recently identified as the gene located in the *IBD1* region that confers susceptibility to CD (Hugot et al 2001, Ogura et al 2001). Three sequence variants (Arg702Trp, Gly908Arg, and Leu1007fsinsC) in this gene were identified by re-sequencing patient DNA samples and were demonstrated to be associated with CD. Importantly, it has recently been shown that common haplotypes constructed with publicly available SNPs detected association to CD with a haplotype block containing the *CARD15* gene (Vermeire et al 2002b). This latter study also demonstrated that these three mutations arose independently on different chromosomes that shared an ancestral haplotype. Taken together, these reports suggest a haplotype-based approach to performing comprehensive association studies of genes or genomic regions (Goldstein 2001). In recognition of the potential of this approach for disease studies, there is now an international effort underway to determine the haplotype structure of the entire human genome (The International HapMap Consortium 2003). This effort, known as the International HapMap Project, will enable future genomewide association studies, and given a sufficiently large sample size, will be a powerful approach for identifying disease susceptibility genes.

Although due to the costs of current technologies it is not feasible to perform comprehensive genomewide association studies using a haplotype-based approach, this approach can now be applied to discrete genomic regions identified by linkage studies or to candidate genes/gene pathways. One example of a genomic region that has been linked to IBD is chromosome 6p13 (known as *IBD3*) that contains the major histocompatibility complex (MHC) region. The MHC region contains over 140 genes, including the eight highly variable classical human leukocyte antigen (HLA) loci. Associations between autoimmune diseases and variation in the MHC region are among the most consistent findings in human genetics. Given the importance of this region, we recently reported a preliminary haplotype map that can be applied to IBD and other autoimmune diseases (Walsh et al 2003). The application of a haplotype-based approach to candidate genes is only limited by our ability to identify relevant genes and pathways to test. Moreover, even if a gene is experimentally implicated in pathogenesis, genetic variation that influences disease may not lie in that gene but rather in another gene in the same biological pathway. Unfortunately we do not have complete knowledge of human biological pathways. However, recent attempts using large sets of expression profiling data have provided new information pertaining to specific biological pathways (Alizadeh et al 2000). It has been suggested that even pair-wise comparisons of gene expression data can suggest novel interactions (although more sophisticated approaches are certainly necessary to uncover complex biological interactions) (Slonim 2002). Another study employed the idea of 'expression neighborhoods' to identify genes with similar expression patterns that may therefore function in common biological pathways. This approach led to the identification of a gene mutated in a mitochondrial disorder (Mootha et al 2003a), and more recently to the identification of a set of genes involved in oxidative phosphorylation whose expression is coordinately decreased in human diabetic tissue (Mootha et al 2003b). In this latter study, no single gene was significantly differentially expressed between diabetic and normal tissue. Only when gene sets were examined using a simple approach called Gene Set Enrichment Analysis (GSEA) was it possible to identify significant differences in gene expression patterns between healthy and diseased tissues. Importantly, all of these strategies enable the exploration of specific biological pathways using general expression data sets (e.g. surveys of normal mouse tissues, surveys of normal human cell lines, etc.), since these methods rely on comparison of gene expression levels across different 'conditions' (different cell types, different stimuli, different developmental stages, etc.). Therefore our ability to explore biological pathways will improve as additional data sets as well as additional integrative approaches are developed.

This brief description of the progress towards identifying IBD disease genes is undoubtedly incomplete, but should demonstrate that there have been many advances in our understanding of the complex genetics that increase a person's susceptibility to developing IBD. In fact, we are beginning to be able to build a model of the genetic interactions leading to IBD and we will soon begin to obtain a better understanding of the interactions between these genetic variants and environmental factors. We still face many challenges, but as we complete the model of the gene–gene and gene–environment interactions we will increase our knowledge of the biological mechanisms underlying the disease processes that result in CD and UC.

References

Abreu MT, Taylor KD, Lin YC et al 2002 Mutations in NOD2 are associated with fibrostenosing disease in patients with Crohn's disease. Gastroenterology 123:679–688

Ahmad T, Armuzzi A, Bunce M et al 2002 The molecular classification of the clinical manifestations of Crohn's disease. Gastroenterology 122:854–866

Alizadeh AA, Eisen MB, Davis RE et al 2000 Distinct types of diffuse large B-cell lymphoma identified by gene expression profiling. Nature 403:503–511

Armuzzi A, Ahmad T, Ling KL et al 2003 Genotype-phenotype analysis of the Crohn's disease susceptibility haplotype on chromosome 5q31. Gut 52:1133–1139

Cavanaugh J 2001 International collaboration provides convincing linkage replication in complex disease through analysis of a large pooled data set: Crohn disease and chromosome 16. Am J Hum Genet 68:1165–1171

Cavanaugh J, Adams KE, Quak EJ et al 2003 CARD15/NOD2 risk alleles in the development of Crohn's disease in the Australian population. Ann Hum Genet 67:35–41

Daly MJ, Rioux JD, Schaffner SF, Hudson TJ, Lander ES 2001 High-resolution haplotype structure in the human genome. Nat Genet 29:229–232

Gabriel SB, Schaffner SF, Nguyen H et al 2002 The structure of haplotype blocks in the human genome. Science 296:2225–2229

Giallourakis C, Stoll M, Miller K et al 2003 IBD5 is a general risk factor for inflammatory bowel disease: replication of association with Crohn disease and identification of a novel association with ulcerative colitis. Am J Hum Genet 73:205–211

Goldstein DB 2001 Islands of linkage disequilibrium. Nat Genet 29:109–111

Hampe J, Cuthbert A, Croucher PJ et al 2001 Association between insertion mutation in NOD2 gene and Crohn's disease in German and British populations. Lancet 357:1925–1928

Hirschhorn JN, Lohmueller K, Byrne E et al 2002 A comprehensive review of genetic association studies. Genet Med 4:45–61

Hugot JP, Chamaillard M, Zouali H et al 2001 Association of NOD2 leucine-rich repeat variants with susceptibility to Crohn's disease. Nature 411:599–603

International HapMap Consortium 2003 The International HapMap Project. Nature 426:789–796

Jeffreys A, Kauppi L, Neumann R 2001 Intensely punctate meiotic recombination in the class II region of the major histocompatibility complex. Nat Genet 29:217–222

Johnson GC, Esposito L, Barratt BJ et al 2001 Haplotype tagging for the identification of common disease genes. Nat Genet 29:233–237

Lander E, Kruglyak L 1995 Genetic dissection of complex traits: guidelines for interpreting and reporting linkage results. Nat Genet 11:241–247

Mirza MM, Fisher SA, King K et al 2003 Genetic evidence for interaction of the 5q31 cytokine locus and the CARD15 gene in Crohn disease. Am J Hum Genet 72:1018–1022

Mootha VK, Lepage P, Miller K et al 2003a Identification of a gene causing human cytochrome c oxidase deficiency by integrative genomics. Proc Natl Acad Sci USA 100:605–610

Mootha VK, Lindgren CM, Eriksson KF et al 2003b PGC-1alpha-responsive genes involved in oxidative phosphorylation are coordinately downregulated in human diabetes. Nat Genet 34:267–273

Negoro K, McGovern DP, Kinouchi Y et al 2003 Analysis of the IBD5 locus and potential gene-gene interactions in Crohn's disease. Gut 52:541–546

Ogura Y, Bonen DK, Inohara N et al 2001 A frameshift mutation in NOD2 associated with susceptibility to Crohn's disease. Nature 411:603–606

Rioux JD, Silverberg MS, Daly MJ et al 2000 Genomewide search in Canadian families with inflammatory bowel disease reveals two novel susceptibility loci. Am J Hum Genet 66:1863–1870

Rioux JD, Daly MJ, Silverberg MS et al 2001 Genetic variation in the 5q31 cytokine gene cluster confers susceptibility to Crohn disease. Nat Genet 29:223–228

Risch N, Merikangas K 1996 The future of genetic studies of complex human diseases. Science 273:1516–1517

Slonim DK 2002 From patterns to pathways: gene expression data analysis comes of age. Nat Genet 32:502–508

Vavassori P, Borgiani P, D'Apice MR et al 2002 3020insC mutation within the NOD2 gene in Crohn's disease: frequency and association with clinical pattern in an Italian population. Dig Liver Dis 34:153

Vermeire S, Louis E, Rutgeerts P et al 2002a NOD2/CARD15 does not influence response to infliximab in Crohn's disease. Gastroenterology 123:106–111

Vermeire S, Wild G, Kocher K et al 2002b CARD15 genetic variation in a Quebec population: prevalence, genotype-phenotype relationship, and haplotype structure. Am J Hum Genet 71:74–83

Walsh EC, Mather KA, Schaffner SF et al 2003 An integrated haplotype map of the human major histocompatibility complex. Am J Hum Genet 73:580–590

DISCUSSION

Jewell: One of the intellectual challenges facing us—while we can sort out haplotype blocks—is that the genes involved are all in such tight linkage disequilibrium and it will be problematic deciding precisely which genes are relevant. We can spend millions of dollars identifying haplotypes, but will we have made any real progress?

Rioux: I agree, performing whole genomewide association studies will not be the end—we are going to identify associated blocks containing multiple genes. To then go from variation to biological effect is the big challenge. To flip this around, it is an important process to go through: rather than trying to guess which genes and pathways within the whole genome have variation which contributes to disease susceptibility, we will have reduced the genome down from 3 billion bases and the 10 million variants to just tens of kilobases and dozens of genes. It's a starting point. The first part of your question concerned how we can do this. If you take *CARD15* as an example, linkage disequilibrium drops off on either side of the gene. This is a highly associated block. It is quite simple, therefore, from a genetic perspective to say it is this gene, but functionally speaking it is still quite difficult because we don't yet know the primary function of this gene. If you bring this to the chromosome 5 or chromosome 6 regions, we have multiple genes that are travelling as a unit. Here, genetics will not bring us down to a single gene. With chromosome 5 we have taken the few hundred families we have studied and have tried to identify rare recombination patterns of this chromosome. We are trying to think of this in an analogous fashion to what we all do in Mendelian disease studies—look for break points. They do exist. Every individual chromosome is a simple mosaic pattern of the different haplotypes across the blocks. If you look at a few hundred families there is an insufficient number of these recombinant forms to refine the association signal because there is very little recombination on the risk haplotype. We have increased this to ~1000 cases and controls and this is still not enough. We are still in the process of modelling this scenario but my guess would be that with 5000–10 000 samples we may have enough recombinant chromosomes in this region to narrow the signal further. Even if we do, we know that there aren't any coding variants that are unique to this risk haplotype. All the other haplotypes are undertransmitted or not overtransmitted. The variant has to be unique to this risk haplotype. Then you would say since it is not in the coding region, it has to affect expression, either splicing, efficiency of splicing or overall expression levels. That doesn't mean it has to be a gene within this haplotype: you have to take a look at a larger region for expression. We already know from the literature that most of these genes are differentially expressed if you compare Crohn's versus normal, for example. This will likely be the same in any region of the genome that we examine. This particular region has recently been shown to be associated with rheumatoid arthritis (Tokuhiro et al 2003) but they only looked at one gene first. They stated that they couldn't break down the association signal and they then started to look at the function of one of these genes. They said that it was immune-specific and differentially expressed, but they could have taken any one of half a dozen other genes in this region and shown the exact same thing. The answer is, I don't know. We are going to have to do knockouts and knock-ins in mice, but what if the genes have different functions in mice? What if the human gene doesn't function well in a mouse model? Take a look at the *Nod2* knockout: what have we learned so far from that? If I had the answer I probably would have given a different talk.

Sartor: The way to solve that dilemma might be to go to partner with the knock-in and knockout strategies. You raised the major issue that *NOD2/CARD15* probably would not have been found if one were using this strategy. This doesn't necessarily say that it's a false hope for other genes.

Rioux: It also has not been fully explored. It is a complex trait in humans: why should we expect it to be any simpler in mice? Knocking out a single gene does not recreate the genetic environment that is in humans.

Sartor: But it does give a means of approaching it scientifically. I would certainly say that in a number of transgenic mice and rats there has been evidence of developing colitis that does not completely recapitulate the human situation but has given us a good enough handle to make some major scientific advances.

Rioux: With the *CARD15* knockout I don't think we have fully explored the knockout mouse. There might be a little more happening. The same thing applies for *IBD5*.

Sartor: One question that might be worth discussing is does it affect strategies if you are looking for a loss of function versus a gain of function? In the knockout mouse in almost all situations I am aware of from the colitis standpoint, you need a homozygous deficiency. There are subtle changes but not a phenotype associated with the single heterozygous knockout. Does that affect the human situation? With *CARD15* it is a 1.5–4-fold relative risk versus a 20–40-fold relative risk for the heterozygous and homozygous loss of function, respectively.

Rioux: The way you do the design of the genetic studies is independent of what you believe to be the functionality downstream. The only effect it has on genetic studies is that the strength of the effect of the alleles will affect the sample size. Something that confers strong risk will be easier to find than something that confers mild risk.

Peña: I am fascinated that the centrepiece of the advances in the genomewide screens is that they are moving towards concentrating on haplotypes. I'd like to come back to the other approach to study genetics of complex diseases, which is the candidate gene association study. As you know a new gene has been found, just outside the 5q31 region, the *TIM1* gene. This gene codes for the receptor of the hepatitis A virus and is also important in the modulation of the Th2 response (McIntire et al 2003). Dr Rioux has shown in the haplotype study of this region in chromosome 5 that there is very little diversity. However, you have also shown that this region is important for the immune response and the regulation of the inflammatory response. It seems that nature has an effective mechanism by assuring a high recombination. Are you not afraid, that if we look into the haplotype only we are probably going to miss the importance of the regulation of the actual genes that we know are important in the regulation of the Th1/Th2 response?

One other comment is that I don't think we should only look to the continents like Africa, North America and Europe in assessing the frequency of the haplotypes. We now know that genes such as *CARD15* and the genes involved in the regulation of the immune response by the intestinal microflora in relatively small homogeneous populations within European countries, for example, show significant differences. In Galicia, Spain the frequency of the commonest *CARD15* mutations are similar to that found in Finland and Scotland (Núñez et al 2004). The frequencies differ significantly from those found in Madrid and other Caucasian population (Mendoza et al 2003). These results also strengthen the variability in frequencies of different genes and haplotypes in homogeneous populations. Have you any comments on this?

Rioux: These are all things that we have been struggling with for a long time. I would do a genomewide association study tomorrow if we had the technology to do it and the ability to do it at a reasonable cost. I wouldn't waste my time looking at pathways. I'd find the associated regions and genes in the genomewide study and then we would have to hammer away at determining the functional mechanisms. I can't do that now, but this may change. All these studies are pushing companies to compete, get better technologies and do things at a much more reasonable cost. If we had talked about sequencing the entire human genome and two dozen other genomes 25 years ago, people would have laughed us out of the room. It is the same thing for whole genome association studies — we are going to be able to do this eventually, and identify the genetic components of the disease. To figure out how they function is a completely different story. But rather than just try to pick our best genes, gene regions or pathways, I would definitely do a genomewide association. The nice thing about this approach is that it doesn't assume that it is a structural difference or a regulatory difference.

To get to your question about important genes regulating other genes, I have no doubt that there are genes which are important for the regulation of genes on chromosome 5, but unless there are genetic variants in those genes, that are common in the population and that have an effect on their function, it doesn't matter. There are lots of important genes in biology, but from a genetics perspective there are only a few that we really care about for any one disease: those that have variants conferring susceptibility. Without a doubt, the human population has gone through certain selective pressures in the past—such pressures are likely to have exerted an effect on specific regions of the genome and this may explain the apparent increase in linkage disequilibrium in regions such as the major histocompatibility complex, the *IBD5* region, and others.

Moore: A year or two ago there was a big article in *Time* magazine about the promise of all these genetic studies and how we are going to be able to cure disease and select patients for appropriate treatment based on their gene profile. Are you saying it is going to cost so much money no one is ever going to do it?

Rioux: I didn't say we are never going to do it, just that it won't be done today.

Moore: What is the future for this approach, relative to the old *in vivo* pharmacology approach?

Rioux: We have to differentiate identifying all the causal genetic variants from identifying pathways that we can interrogate or modulate pharmacologically. Both approaches can have a huge impact on disease. The genetics is still just only a part of all this. Certainly, understanding the pathways which we know are important for the immune response to bacteria is important, especially if we can then modulate them to affect disease course or response to therapy. Those are still good approaches. But as a geneticist, over the next couple of years I am still going to be looking at biological pathways to look for genetic variants. As I have made clear, the way to go in two or three years time will be to do genomewide association, because we will be able to narrow down the genome to say just 12 genes that we need to target.

Schölmerich: With respect to Salvador Peña's earlier comment we should be careful that we do not follow the 'Casablanca approach', which is to arrest the usual suspects. We all are bound by this immunological paradigm and we forget that the barrier is much more than immunology: the barrier is metabolism, tight junctions and cadherins and much more. Genetic changes of these will only be found using the systematic approach that John Rioux has described. We need both sides. If we follow only one paradigm and the whole community does this, then you run into a dead end street. We need the geneticists doing association studies and telling us that it could be something completely different, such as lipid metabolism. Lipids are actually very important in detoxifying bacterial products. It doesn't have to be just interleukins. I'm getting a bit tired with interleukins!

Sartor: None of the colitis models have come out of a targeted approach aimed at making a colitis model. All have come out of serendipity: someone has knocked out something because they are interested in the pathway and the mouse gets diarrhoea and loses weight. Serendipity rules.

Rhodes: I am always surprised that when you and your fellow geneticists present these talks there is no mention of correction for environmental factors, in particular smoking status. There is also no mention of correction for age. Might you be missing important associations?

Rioux: As geneticists, more often than not we ignore environment. It is not because we want to, but because currently most of our collections don't have reliable information about environment. One of the things we need to think about for ongoing and future collections is to have better environmental data. Having said that, we know a bit about the effect of smoking but we know about very few other environmental factors to examine. We need to pass the buck to the epidemiologists.

Rhodes: I think the buck stops with you! Smoking is a powerful factor and it shouldn't be too difficult to correct for it.

Rioux: It comes back to a single hypothesis. We can definitely examine smoking if we do a better job at collecting these data. To be fair, however, we wouldn't want to limit ourselves to that alone. Why would we spend thousands of dollars to do a new genetic collection and collect just smoking as an environmental factor? This would seem like a waste of effort and money. We need to put more effort into identifying additional environmental factors that we would need to include in the genetic studies and for this I believe that we need better epidemiological and natural history studies.

Jewell: It is clearly an important point and there is an increasing amount of phenotypic and environmental data being collected in all the centres.

Rhodes: Is smoking a risk factor for developing Crohn's disease irrespective of an individual's *CARD15* genotype?

Ahmad: In Oxford we have recently investigated possible gene-environmental interactions in Crohn's disease (CD) by examining the influence of smoking on disease susceptibility in individuals at high genetic risk. We recruited 433 discordant sib pairs (DSPs) from 289 families (351 CD patients and 365 unaffected siblings). One DSP was selected from each family. In families with multiple DSPs the eldest unaffected and youngest CD sibs were selected. Prolonged follow up of unaffected siblings (15.8 years longer than the age of diagnosis of their affected sibling) minimised risk of diagnostic misclassification. CD patients were defined as smokers if they smoked prior to diagnosis; in the absence of a fixed time point, unaffected siblings were defined as smokers if they had ever smoked. Overall, siblings demonstrated 68% concordance for smoking history and there was no difference in the prevalence of smoking in affected and unaffected sibs. However in individuals with a high genetic risk (two *NOD2* variant alleles), 0/6 unaffected siblings were smokers compared with 11/24 CD patients ($P < 0.05$). This small study is limited by the fact that sib-pairs are over-matched for environmental factors including smoking. Nonetheless, these data suggest that smoking may act as an important environmental modifier of disease susceptibility in genetically high-risk individuals.

References

McIntire JJ, Umetsu SE, Macaubas C et al 2003 Immunology: hepatitis A virus link to atopic disease. Nature 425:576

Mendoza JL, Murillo LS, Fernandez L et al 2003 Prevalence of mutations of the NOD2/CARD15 gene and relation to phenotype in Spanish patients with Crohn disease. Scand J Gastroenterol 38:1235–1240

Núñez C, Barreiro M, Dominguez-Munoz JE, Lorenzo A, Zapata C, Pena AS 2004 CARD15 mutations in patients with Crohn's disease in a homogeneous Spanish population. Am J Gastroenterol 99:450–456

Tokuhiro S, Yamada R, Chang X et al 2003 An intronic SNP in a RUNX1 binding site of SLC22A4, encoding an organic cation transporter, is associated with rheumatoid arthritis. Nat Genet 35:341–348

Phenotype-determining genes in inflammatory bowel disease

T. Ahmad

Gastroenterology Unit, University of Oxford, Gibson Laboratories, Radcliffe Infirmary, Woodstock Road, Oxford OX2 6QX, UK

Abstract. Inflammatory bowel disease (IBD) has traditionally been categorized as either ulcerative colitis or Crohn's disease on the basis of clinical, radiological and histological criteria. Within these diseases however, significant heterogeneity is observed, suggesting the existence of phenotypic subtypes, based on features such as location and behaviour of disease. Evidence for a possible genetic basis for these subgroups emerged in the 1990s from epidemiological studies in multiply affected families. Recent advances in our understanding of the relationship between genotype and phenotype in IBD now challenge traditional clinical classifications, promising a taxonomy of disease based on precise molecular, rather than ambiguous clinical definitions. While many of the genes remain unidentified, the emerging data suggest that IBD comprises a heterogeneous family of oligogenic inflammatory disorders in which the specific clinical manifestations of disease in any individual are determined by the interaction of genetic and environmental factors. These data have highlighted both the importance and difficulties of classifying patients into accurately defined clinical subgroups, and suggest that a genetic basis for the observed disease heterogeneity may account for the discrepant findings from earlier genetic studies. Despite the considerable insights offered to investigators these advances have yet to impact on the clinical management of patients.

2004 Inflammatory bowel disease — crossroads of microbes, epithelium and immune systems. Wiley, Chichester (Novartis Foundation Symposium 263) p 17–40

Diagnosis and classification of disease is central to the practice of medicine, permitting the identification of subgroups that differ in aetiology, natural history, prognosis and response to therapy. Inflammatory bowel disease (IBD) has traditionally been categorized as either ulcerative colitis (UC) or Crohn's disease (CD) on the basis of clinical, radiological and histological criteria. However over the last 30 years the heterogeneous nature of these diseases has become increasingly apparent. These clinical observations have suggested the existence of phenotypic subtypes based on features such as location, behaviour, natural history and response to treatment. Recent advances in our understanding of the genetics of IBD, in particular the identification of *NOD2/CARD15*, have provided the opportunity to explore the genetic basis for this clinical heterogeneity. This presentation reviews recent progress in our understanding of the contribution of genetics to IBD phenotype and considers its implications for clinical practice.

The clinical heterogeneity of the inflammatory bowel diseases

The first attempts to classify patients with CD were made in the 1970s based on the anatomical location of disease. Data from subsequent studies validated this approach by demonstrating that disease location has important implications for both medical (Farmer et al 1975, Franchimont et al 1998) and surgical therapy (Farmer et al 1985, Whelan et al 1985, Griffiths et al 1991). In 1988 additional distinct subgroups of CD were described, defined by disease behaviour (Greenstein et al 1988). Perforating disease was shown to be distinct from non-perforating disease, a finding that has been confirmed in subsequent studies (Aeberhard et al 1996). Moreover, recurrent postoperative disease was shown to follow the same behaviour pattern as the primary disease (Greenstein et al 1988). Data from these observational studies led to the development of a formal consensus classification of CD, with the objective of standardizing the description of study populations in clinical trials, and to aid the correlation of putative aetiological factors with particular clinical phenotypes. The first such formal attempt was produced in 1992 (Sachar et al 1992) and subsequently refined in 1998 to produce the Vienna classification (Gasche et al 2000). In support of this clinical classification of disease, anti-*Saccharomyces cerevisiae* antibodies (ASCA) and anti-neutrophil cytoplasmic antibodies (ANCA) have been shown to define discrete subgroups of CD: pANCA expression in CD defining a left-sided colitis similar to UC (Vasiliauskas et al 1996).

Epidemiological evidence for a genetic contribution to the disease heterogeneity of IBD

From the mid 1990s evidence for a possible genetic basis for these subgroups emerged from epidemiological studies in multiply affected families (Bayless et al 1996, Colombel et al 1996, Peeters et al 1996, Polito et al 1996, Satsangi et al 1996a, Annese et al 2001). The evidence is particularly convincing for CD, which has been more extensively studied than UC. In a study from Oxford 82% of affected siblings were concordant for disease type (CD or UC), 76% for disease extent and 84% for extra-intestinal manifestations. Concordance was greater in siblings than in parent–child pairs (Satsangi et al 1996a). A similar study from Johns Hopkins of 60 multiply-affected CD families identified concordance rates of 86% for disease site and 82% for disease behaviour (Bayless et al 1996). Similar findings have been reported from centres in France (Colombel et al 1996) and Belgium (Peeters et al 1996). Of note, concordance rates in the French study increased with the number of affected relatives within the family (Colombel et al 1996). Fewer data on familial concordance are available for UC. In a study from Oxford concordance rates for disease extent were 53% in 17 parent–child pairs and 69% in 35 sibling pairs (Satsangi et al 1996a). These figures were lower than for CD–CD pairs, consistent with other evidence suggesting a smaller contribution of genetics to disease susceptibility in UC compared to CD.

Design of genotype–phenotype studies

The essential requirement for any study investigating the genetic basis for disease phenotype is a large, accurately characterized cohort of patients. Even low rates of diagnostic misclassification can lead to significant loss of power (Silverberg et al 2001). The ideal study will categorize patients using clearly defined phenotypic variables that demonstrate good inter- and intra-observer variability. Classification should preferably be applicable at diagnosis and stable with time, and all findings should be replicated in a second cohort. When interpreting the data from genotype-phenotype studies it should be remembered that these standards are difficult to meet in IBD. There are a number of reasons for this. Firstly existing internationally agreed classification systems are insufficiently detailed to prevent ambiguous assignment of disease subgroup. This is exemplified by the difficulties in definition and classification of perianal disease; does a skin tag constitute perianal disease? Should perianal and internal fistulating disease be considered together as 'fistulating disease'? Not surprisingly, poor inter- and intra-observer variability may result, even amongst experts (Steinhart et al 1998), which is likely to be a particular problem when using multicentre cohorts. Secondly classifying clinical subgroups may be difficult with phenotypes which are not stable, such as CD behaviour. In a Belgian study, although location of disease remained relatively stable, there were striking changes in disease behaviour over time (Louis et al 2001). Similar problems are encountered when phenotyping UC, as disease may extend or regress unpredictably with time (Powell-Tuck et al 1977, Moum et al 1999). To minimize these limitations disease subtype should not be classified at diagnosis or at a single point in time, but only after a number of years of follow-up.

Adequate sample size is crucial to ensure studies are appropriately powered. This is important in any complex disease study challenged by genetic heterogeneity, epistatic interactions and loci that confer small relative risks. It is however particularly important in genotype–phenotype association studies where multiple subgroup analyses lead to small underpowered subgroups and a substantial risk of associations being described by chance (type 1 error). In order to minimize these risks, positive findings should be replicated in an independent cohort. Publication bias of positive results further contributes to an expanding literature, which is teeming with associations that cannot be replicated.

Genetic determinants of IBD phenotype *NOD2/CARD15*

Association with location and behaviour of Crohn's disease

The most convincing evidence for a genotype–phenotype association in IBD is illustrated by recent studies of the *NOD2/CARD15* gene. As of September 2003, 10 published studies have examined the phenotypic expression of the *NOD2/CARD15* variants. Phenotype definitions, methods of classification and duration of follow-up have varied widely. Nonetheless all but one of these studies have reported association with either ileal (Ahmad et al 2002a, Cuthbert et al 2002, Hampe et al 2002, Murillo et al 2002, Vermeire et al 2002) or fibrostenosing disease (Abreu et al 2002, Lesage et al 2002, Radlmayr et al 2002), or both (Vavassori et al 2002). Dissecting out whether the primary association is with ileal rather than stenosing disease is difficult as these clinical variables are not independent. These associations are present in Jewish (Abreu et al 2002) and non-Jewish populations, and are apparent with all three of the *NOD2/CARD15* variants, although in two studies the relative risk was greatest with the frameshift mutation (Abreu et al 2002, Ahmad et al 2002a). Interestingly the association also appears to be present in patients with ileo-colonic disease, suggesting that the *NOD2/CARD15* variants are associated with ileal disease irrespective of the presence of colonic disease. This unprecedented concordance between centres is particularly significant considering the difficulties of phenotype definition discussed above. However, it is important to note that not all patients with ileal disease have *NOD2/CARD15* mutations, suggesting that mutations in other genes (or other *NOD2* mutations) might produce a similar phenotype. The variant alleles do not appear to determine the severity of ileal or fibrostenosing disease, as judged by the time to first surgery (Brant et al 2003) or surgical recurrence (Ahmad et al 2002a).

It is not yet clear how *NOD2/CARD15* variants cause ileal or fibrostenosing disease but the recent demonstration of *NOD2/CARD15* expression by Paneth cells, which are located predominately in the terminal ileum provides a tantalizing clue.

Association with familial disease?

Several epidemiological studies have reported a higher incidence of ileal or ileocolonic disease in familial cases of CD (Colombel et al 1996, Cottone et al 1997, Halme et al 2002). Seven studies analysed association between family history and *NOD2/CARD15* status, of which three reported significantly higher frequencies of *NOD2/CARD15* polymorphism in patients with familial disease (Cuthbert et al 2002, Vavassori et al 2002, Zhou et al 2002). Of note in the study from New York, comprising 481 exclusively Jewish patients, association with the *NOD2* variants was only present in patients with familial disease (Zhou et al 2002).

Association with early age at diagnosis

Five studies analysed association between age of onset and *NOD2/CARD15* status, of which three reported association between early age of diagnosis and possession of two variant alleles (Ahmad et al 2002a, Lesage et al 2002, Zhou et al 2002).

IBD5 and CD

In 2001, linkage disequilibrium mapping of the *IBD5* locus on chromosome 5 identified a common, highly conserved 250 kb haplotype associated with CD (Rioux et al 2001). Both the degree of linkage disequilibrium and the density of immunoregulatory genes within this locus have hampered the identification of the causal mutation on this haplotype. In a genotype–phenotype study from Oxford the risk haplotype is shown to be specifically associated with perianal CD, with the greatest relative risk seen in individuals homozygous for the risk haplotype (RR 3–4) (Armuzzi et al 2003).

HLA complex and IBD

The HLA region (encompassed within *IBD3*) is a highly complex, gene dense region located on chromosome 6p. It encodes the classical class I and II genes essential for normal lymphocyte function, and in addition, a further 200+ genes, many of which have immunoregulatory roles. While there can be no doubt that genes in the HLA region contribute to IBD susceptibility, the majority of reported associations in phenotypically unselected CD and UC cohorts have been inconsistent, even within defined ethnic groups, reflecting many of the methodological problems discussed earlier. More convincing associations with discrete phenotypic subgroups have emerged from recent studies, suggesting that perhaps genes in the HLA region may have a greater role in modifying phenotype than determining disease susceptibility.

HLA and UC

In UC, the most convincing HLA associations have been with Class II alleles and disease susceptibility. To date, the most consistent associations have been between DRB1*0103, DRB1*1502, and DRB1*0401, all of which have been highlighted by a recent meta-analysis (Stokkers et al 1999). Within UC subgroups, the uncommon DRB1*0103 is consistently associated with both extensive disease and severe disease requiring colectomy (Satsangi et al 1996b, Roussomoustakaki et al 1997, Bouma et al 1999, Ahmad et al 2002b, Yamamoto-Furusho et al 2003). Amongst patients requiring colectomy carriage of DRB1*0103 is also associated with shorter time to surgery (9 months vs. 5.2 years $P=0.002$). However the low frequency of this allele, even in the UC population, suggests that this association is unlikely to be clinically useful in predicting disease course.

TNF — modifier of disease extent in UC?

A potentially interesting new finding is the association of homozygosity for the common tumour necrosis factor (TNF) promoter haplotype (TNF −1031T −863C −857C −380G −308G −238G) and the presence of limited UC that remains distal over time (Ahmad et al 2002b). At present this association should be interpreted with extreme caution in view of both the difficulties of defining disease extent in UC, and lack of independent replication. TNFα is an important proinflammatory cytokine, and has been implicated in the pathogenesis of a number of inflammatory disorders. Polymorphisms at positions −1031, −863, −857 and −308 have been associated with increased transcriptional activity and production of TNFα in some, but not all studies, reflecting the range of experimental techniques used. These results might suggest that UC limited to the distal colon, is associated with a low TNF-producing haplotype. Thus TNF may act as a modifying gene in UC, acting to alter phenotypic expression of disease, whilst exerting no detectable effect on overall susceptibility.

HLA: susceptibility locus for colonic CD

Interestingly, DRB1*0103, which has been associated with both UC susceptibility and clinical subtype, has also been associated with CD (Trachtenberg et al 2000, Ahmad et al 2002a, Silverberg et al 2003). In a recent Canadian study, subgroup analysis revealed that an association with DRB1*0103 was observed only in CD patients with pure colonic disease (Silverberg et al 2003). Although this study is underpowered, the observation of a shared association of DRB1*0103 with UC and colonic CD provides a tantalising clue as to the potential molecular basis for definition of this subgroup. Furthermore, the existence of a shared UC and CD susceptibility allele on chromosome 6 may explain why both forms of IBD can coexist in a family at a frequency greater than expected by chance. Indeed, whereas *NOD2/CARD15* is strongly associated with ileal CD, the HLA region is particularly associated with colonic CD. In addition to the association of colonic disease with the rare DRB1*0103 allele, we have recently shown an association with the common classical autoimmune haplotype (A1B8DR3) and colonic CD, with or without ileal disease, with a relative risk of 3 (Ahmad et al 2002a). While this requires replication, these data are consistent with reports from a Belgian group demonstrating association of colonic disease with the TNF promoter allele −308A, which occurs on this haplotype (Louis et al 2000, 2002).

HLA genes and the extra-intestinal manifestations of IBD

Phenotype may also be defined by the presence of extra-intestinal manifestations of IBD and although inevitably limited by inadequate sample size a number of genetic associations have been described: in studies comprising both UC and CD patients, a migratory pauci-articular large joint arthritis was shown to be associated with HLA-DRB1*0103, B*27 and B*35, whereas a chronic, small joint, symmetrical arthritis was associated with HLA-B*44 (Orchard et al 2000). Uveitis has been associated with HLA-B*27 and DRB1*0103 and erythema nodosum with the TNF promoter SNP TNF-1031C (Orchard et al 2002).

IL1RA and UC

Outside of the HLA region, the interleukin 1 (*IL1*) gene family on chromosome 2q13 is the most extensively studied group of genes in IBD. An association with allele 2 of a variable tandem repeat sequence (VNTR) in intron 2 of the *IL1RA* has remained highly controversial since its original report in 1994. The last 18 months has seen the publication of three adequately powered studies comprising 529 (347 UC) (Craggs et al 2001), 320 (all UC) (Carter et al 2001a) and 342 (124 UC) (Vijgen et al 2002) IBD patients. Interestingly, the two studies from the North of England produced conflicting data: the Newcastle group found no association of this VNTR with either susceptibility or clinical subgroup in CD or UC (Craggs et al 2001). In contrast, the group from Sheffield, authors of the original report in 1994, continue to find a weak association with susceptibility to UC (OR 1.3), most marked in those patients with extensive disease (OR 1.5) or those undergoing colectomy (OR 1.5) (Carter et al 2001a). Further analysis of 82 UC patients from this same cohort, who had undergone colectomy and ileal pouch-anal anastomosis, demonstrated that the association with this allele was strongest in those who developed pouchitis (relative hazard 3.1) (Carter et al 2001b). In the most recent study from Belgium, a significant decrease in allele 1 homozygosity was identified, but no association with allele 2 (Vijgen et al 2002). In summary, the association with this intronic VNTR in the *IL1RA* gene is at best weak.

MDR1 and UC

MDR1 encodes Pgp, which plays a major role in epithelial cell transport and barrier function. It is an attractive functional candidate gene as it is located within an area of linkage on chromosome 7, and the immunologically normal *Mdr1a*$^{-/-}$ knockout mouse develops spontaneous colitis, which can be prevented with oral antibiotics. A weak association of UC with a functionally significant SNP C3435T, in exon 26 of the *MDR1* gene (Schwab et al 2003), has recently been reported in a study of 149 patients with UC. However, in a similar study of 317 UC patients this SNP was not associated with overall susceptibility or the need for surgery, demonstrating the importance of replication (D. McGovern, personal communication).

Genetic determinants of bone loss in IBD

Bone loss and osteoporosis are recognized sequelae of IBD, but the risk factors have not been clearly identified. Two small studies have recently investigated whether genetic markers may predict bone loss. In the first study of 83 patients, associations with bone loss were identified with polymorphisms in *IL6* and *IL1RA* (Schulte et al 2000), genes previously implicated in the paracrine stimulation of osteoclast development and resorption of bone. In the second study comprising 75 patients association was reported with an *IL1β* polymorphism (Nemetz et al 2001), previously associated with increased *in vitro* production of *IL1β* by mononuclear cells. Both of these studies were limited by small underpowered, heterogeneous study populations, and a failure to correct for multiple comparisons. Larger studies are now required to replicate these data.

Predicting response to drug therapy

IBD phenotype may also be defined on the basis of drug response, which itself may be genetically determined. This emerging field of pharmacogenomics is outside the scope of this paper.

Conclusions

For 30 years data from genetic studies in IBD have been disappointingly inconsistent. The recent identification of *NOD2/CARD15* represents a major landmark in our understanding of IBD genetics. Unlike previously reported genetic associations, the biological plausibility, strength of association, dose–response effect and consistency between independent studies, suggest a causal role for the *NOD2/CARD15* variants. Subsequent reports demonstrating the specific association with ileal, or fibrostenosing CD only, have validated the

TABLE 1 Future applications from genotype–phenotype studies

Aetiology

- Facilitate understanding of molecular mechanisms specific to disease subgroups
- Enable identification of environmental factors specific to disease subgroups
- Provide insight into the aetiology of other disorders
- Permit accurate early diagnosis based upon molecular mechanisms, rather than phenotype

Screening of 'at risk' individuals

- Target prevention at individuals at risk

Predict disease course

- Need for surgery
- Post-operative recurrence
- Development of cancer
- Development of extra-intestinal manifestations
- Development of osteoporosis

Therapeutics

- Drug development focused on specific molecularly defined disease subtypes
- Ensure disease homogeneity in pharmaceutical studies
- Optimal patient treatment tailored to specific disease subtype
- Prediction of adverse drug reactions

approach of classifying patients into accurately defined clinical subgroups. This raises the possibility that a genetic basis for the observed disease heterogeneity in both UC and CD may account for the discrepant findings from earlier genetic studies in unselected populations.

Understanding the genetic basis for IBD phenotype represents the crucial first step in understanding the molecular mechanisms specific to subgroups of disease. The emerging genetic data suggest that IBD comprises a heterogeneous family of oligogenic inflammatory disorders in which the specific clinical manifestations of disease in any individual are determined by the interaction of a number of genetic variants and environmental factors. Demonstration that not all patients with ileal CD have *NOD2/CARD15* variants suggests that the same phenotype may be caused by mutations in different genes. Such genetic heterogeneity, which may occur within homogeneous populations, is likely to be most marked between different ethnic groups. Mutations in the same gene may confer susceptibility to more than one clinically defined disorder. This is demonstrated by the shared association of DRB1*0103 with both colonic CD and ulcerative colitis. In addition this shared susceptibility may extend to other inflammatory disorders, previously not considered to be related to IBD. This may explain the coexistence of two diseases in the same family.

Data from recent genotype–phenotype studies are already being used in the laboratory to redefine disease on a molecular basis. This has immediate scientific application through the stratification of genetic studies, necessary given the extent of locus heterogeneity in IBD. Despite these advances, current knowledge of the genetic basis for IBD phenotype is of little clinical value in the diagnosis, classification (e.g. indeterminate colitis) or prediction of disease course. For the time being diagnosis and management will continue to be a clinical process, but as more is learnt about the genetic basis for individual subgroups, clinical applications will follow (Table 1).

Acknowledgements

Tariq Ahmad is supported by grants from the National Association of Crohn's and Colitis (NACC, UK) and the Broad Foundation (USA).

References

Abreu MT, Taylor KD, Lin YC et al 2002 Mutations in NOD2 are associated with fibrostenosing disease in patients with Crohn's disease. Gastroenterology 123:679–688

Aeberhard P, Berchtold W, Riedtmann HJ, Stadelmann G 1996 Surgical recurrence of perforating and nonperforating Crohn's disease. A study of 101 surgically treated patients. Dis Colon Rectum 39:80–87

Ahmad T, Armuzzi A, Bunce M et al 2002a The molecular classification of the clinical manifestations of Crohn's disease. Gastroenterology 122:854–866

Ahmad T, Armuzzi A, Bunce M et al 2002b Role of the HLA in determining susceptibility and phenotype in Caucasian ulcerative colitis. Gastroenterology 122:A302

Annese V, Andreoli A, Astegiano M et al 2001 Clinical features in familial cases of Crohn's disease and ulcerative colitis in Italy: a GISC study. Italian Study Group for the Disease of Colon and Rectum. Am J Gastroenterol 96:2939–2945

Armuzzi A, Ahmad T, Ling KL et al 2003 Genotype-phenotype analysis of the Crohn's disease susceptibility haplotype on chromosome 5q31. Gut 52:1133–1139

Bayless TM, Tokayer AZ, Polito JM 2nd, Quaskey SA, Mellits ED, Harris ML 1996 Crohn's disease: concordance for site and clinical type in affected family members—potential hereditary influences. Gastroenterology 111:573–579

Bouma G, Crusius JB, Garcia-Gonzalez MA et al 1999 Genetic markers in clinically well defined patients with ulcerative colitis (UC). Clin Exp Immunol 115:294–300

Brant SR, Picco MF, Achkar JP et al 2003 Defining complex contributions of NOD2/CARD15 gene mutations, age at onset, and tobacco use on Crohn's disease phenotypes. Inflamm Bowel Dis 9:281–289

Carter MJ, di Giovine FS, Jones S et al 2001a Association of the interleukin 1 receptor antagonist gene with ulcerative colitis in Northern European Caucasians. Gut 48:461–467

Carter MJ, Di Giovine FS, Cox A et al 2001b The interleukin 1 receptor antagonist gene allele 2 as a predictor of pouchitis following colectomy and IPAA in ulcerative colitis. Gastroenterology 121:805–811

Colombel JF, Grandbastien B, Gower-Rousseau C et al 1996 Clinical characteristics of Crohn's disease in 72 families. Gastroenterology 111:604–607

Cottone M, Brignola C, Rosselli M et al 1997 Relationship between site of disease and familial occurrence in Crohn's disease. Dig Dis Sci 42:129–132

Craggs A, West S, Curtis A et al 2001 Absence of a genetic association between IL-1RN and IL-1B gene polymorphisms in ulcerative colitis and Crohn disease in multiple populations from northeast England. Scand J Gastroenterol 36:1173–1178

Cuthbert AP, Fisher SA, Mirza MM et al 2002 The contribution of NOD2 gene mutations to the risk and site of disease in inflammatory bowel disease. Gastroenterology 122:867–874

Farmer RG, Hawk WA, Turnbull RB Jr 1975 Clinical patterns in Crohn's disease: a statistical study of 615 cases. Gastroenterology 68:627–635

Farmer RG, Whelan G, Fazio VW 1985 Long-term follow-up of patients with Crohn's disease. Relationship between the clinical pattern and prognosis. Gastroenterology 88: 1818–1825

Franchimont DP, Louis E, Croes F, Belaiche J 1998 Clinical pattern of corticosteroid dependent Crohn's disease. Eur J Gastroenterol Hepatol 10:821–825

Gasche C, Scholmerich J, Brynskov J et al 2000 A simple classification of Crohn's disease: report of the Working Party for the World Congresses of Gastroenterology, Vienna 1998. Inflamm Bowel Dis 6:8–15

Greenstein AJ, Lachman P, Sachar DB et al 1988 Perforating and non-perforating indications for repeated operations in Crohn's disease: evidence for two clinical forms. Gut 29:588–592

Griffiths AM, Wesson DE, Shandling B, Corey M, Sherman PM 1991 Factors influencing postoperative recurrence of Crohn's disease in childhood. Gut 32:491–495

Halme L, Turunen U, Helio T et al 2002 Familial and sporadic inflammatory bowel disease: comparison of clinical features and serological markers in a genetically homogeneous population. Scand J Gastroenterol 37:692–698

Hampe J, Grebe J, Nikolaus S et al 2002 Association of NOD2 (CARD 15) genotype with clinical course of Crohn's disease: a cohort study. Lancet 359:1661–1665

Lesage S, Zouali H, Cezard JP et al 2002 CARD15/NOD2 mutational analysis and genotype-phenotype correlation in 612 patients with inflammatory bowel disease. Am J Hum Genet 70:845–857

Louis E, Peeters M, Franchimont D et al 2000 Tumour necrosis factor (TNF) gene polymorphism in Crohn's disease (CD): influence on disease behaviour? Clin Exp Immunol 119:64–68

Louis E, Collard A, Oger AF, Degroote E, Aboul Nasr El Yafi FA, Belaiche J 2001 Behaviour of Crohn's disease according to the Vienna classification: changing pattern over the course of the disease. Gut 49:777–782

Louis E, Vermeire S, Rutgeerts P et al 2002 A positive response to infliximab in Crohn disease: association with a higher systemic inflammation before treatment but not with -308 TNF gene polymorphism. Scand J Gastroenterol 37:818–824

Moum B, Ekbom A, Vatn MH, Elgjo K 1999 Change in the extent of colonoscopic and histological involvement in ulcerative colitis over time. Am J Gastroenterol 94: 1564–1569

Murillo L, Crusius JB, van Bodegraven AA, Alizadeh BZ, Pena AS 2002 CARD15 gene and the classification of Crohn's disease. Immunogenetics 54:59–61

Nemetz A, Toth M, Garcia-Gonzalez MA et al 2001 Allelic variation at the interleukin 1beta gene is associated with decreased bone mass in patients with inflammatory bowel diseases. Gut 49:644–469

Orchard TR, Thiyagaraja S, Welsh KI, Wordsworth BP, Hill Gaston JS, Jewell DP 2000 Clinical phenotype is related to HLA genotype in the peripheral arthropathies of inflammatory bowel disease. Gastroenterology 118:274–278

Orchard TR, Chua CN, Ahmad T, Cheng H, Welsh KI, Jewell DP 2002 Uveitis and erythema nodosum in inflammatory bowel disease: clinical features and the role of HLA genes. Gastroenterology 123:714–718

Peeters M, Nevens H, Baert F et al 1996 Familial aggregation in Crohn's disease: increased age-adjusted risk and concordance in clinical characteristics. Gastroenterology 111: 597–603

Polito JM 2nd, Childs B, Mellits ED, Tokayer AZ, Harris ML, Bayless TM 1996 Crohn's disease: influence of age at diagnosis on site and clinical type of disease. Gastroenterology 111:580–586

Powell-Tuck J, Ritchie JK, Lennard-Jones JE 1977 The prognosis of idiopathic proctitis. Scand J Gastroenterol 12:727–732

Radlmayr M, Torok HP, Martin K, Folwaczny C 2002 The c-insertion mutation of the NOD2 gene is associated with fistulizing and fibrostenotic phenotypes in Crohn's disease. Gastroenterology 122:2091–2092

Rioux JD, Daly MJ, Silverberg MS et al 2001 Genetic variation in the 5q31 cytokine gene cluster confers susceptibility to Crohn disease. Nat Genet 29:223–228

Roussomoustakaki M, Satsangi J, Welsh K et al 1997 Genetic markers may predict disease behavior in patients with ulcerative colitis. Gastroenterology 112:1845–1853

Sachar DB, Andrews HA, Farmer RG et al 1992 Proposed classification of patient subgroups in Crohn's disease. Gastroenterol Int 5:141–154

Satsangi J, Grootscholten C, Holt H, Jewell DP 1996a Clinical patterns of familial inflammatory bowel disease. Gut 38:738–741

Satsangi J, Welsh KI, Bunce M et al 1996b Contribution of genes of the major histocompatibility complex to susceptibility and disease phenotype in inflammatory bowel disease. Lancet 347:1212–1217

Schulte CM, Dignass AU, Goebell H, Roher HD, Schulte KM 2000 Genetic factors determine extent of bone loss in inflammatory bowel disease. Gastroenterology 119: 909–920

Schwab M, Schaeffeler E, Marx C et al 2003 Association between the C3435T MDR1 gene polymorphism and susceptibility for ulcerative colitis. Gastroenterology 124:26–33

Silverberg MS, Daly MJ, Moskovitz DN et al 2001 Diagnostic misclassification reduces the ability to detect linkage in inflammatory bowel disease genetic studies. Gut 49:773–776

Silverberg M, Mirea L, Bull S et al 2003 A population- and family-based study of Canadian families reveals association of HLA-DRB1*0103 with colonic involvement in inflammatory bowel disease. Inflamm Bowel Dis 9:1–9

Steinhart AH, Girgrah N, McLeod RS 1998 Reliability of a Crohn's disease clinical classification scheme based on disease behavior. Inflamm Bowel Dis 4:228–234

Stokkers PC, Reitsma PH, Tytgat GN, van Deventer SJ 1999 HLA-DR and -DQ phenotypes in inflammatory bowel disease: a meta-analysis. Gut 45:395–401

Trachtenberg EA, Yang H, Hayes E et al 2000 HLA class II haplotype associations with inflammatory bowel disease in Jewish (Ashkenazi) and non-Jewish caucasian populations. Hum Immunol 61:326–333

Vasiliauskas EA, Plevy SE, Landers CJ et al 1996 Perinuclear antineutrophil cytoplasmic antibodies in patients with Crohn's disease define a clinical subgroup. Gastroenterology 110:1810–1819

Vavassori P, Borgiani P, D'Apice MR et al 2002 3020insC mutation within the NOD2 gene in Crohn's disease: frequency and association with clinical pattern in an Italian population. Dig Liver Dis 34:153

Vermeire S, Wild G, Kocher K et al 2002 CARD15 genetic variation in a Quebec population: prevalence, genotype-phenotype relationship, and haplotype structure. Am J Hum Genet 71:74–83

Vijgen L, Van Gysel M, Rector A et al 2002 Interleukin-1 receptor antagonist VNTR-polymorphism in inflammatory bowel disease. Genes Immun 3:400–406

Whelan G, Farmer RG, Fazio VW, Goormastic M 1985 Recurrence after surgery in Crohn's disease. Relationship to location of disease (clinical pattern) and surgical indication. Gastroenterology 88:1826–1833

Yamamoto-Furusho JK, Uscanga LF, Vargas-Alarcon G et al 2003 Clinical and genetic heterogeneity in Mexican patients with ulcerative colitis. Hum Immunol 64:119–123

Zhou Z, Lin XY, Akolkar PN et al 2002 Variation at NOD2/CARD15 in familial and sporadic cases of Crohn's disease in the Ashkenazi Jewish population. Am J Gastroenterol 97: 3095–3101

DISCUSSION

Jewell: You came out very much in favour of regional enteritis being a distinct disease. We could get into a philosophical discussion about disease classification. You suggested the pattern of disease and behaviour of disease is very much an individual matter because it is dependent on your genetic make up. For example, if you have HLA DRB 103 and you have ulcerative colitis you are much more likely to have an aggressive disease leading to early colectomy than if you have DRB1*0301, for example. If you have that allele (DRB103) you are likely to get an acute joint or uveitis. If there was one environmental factor and extensive genetic variation (which we know there is) then you are going to get multiple phenotypes. Is that the same disease or different diseases? The problem then becomes that there is not a single environmental factor but a number of them. It is then very difficult to say that it is a single disease, or back to what we have said for many years over a glass of port — a spectrum of diseases.

Ahmad: That is an extremely difficult question to answer and depends upon how one chooses to classify disease. Certainly, the emerging genetic data suggest that IBD comprises a heterogeneous family of oligogenic inflammatory disorders. Classifying IBD into discrete molecular entities is an attractive concept which is already being exploited by researchers to reduce the genetic and clinical heterogeneity of IBD studies. However it is clear that the specific clinical manifestations of disease in any individual are determined by the complex interaction of genetic and environmental factors, many of which are currently unknown. Such a complex interplay of factors produces the apparent spectrum of disease in IBD making accurate classification into discrete groups considerably more difficult than we have suggested.

Sartor: You commented on the genetic–bacterial interaction and presented the idea that different bacteria can influence the phenotype of disease. Studies in mice have demonstrated that two different bacteria may produce different disease phenotypes in a set genetic background: IL10 knockout mice raised in a germ-free environment do not develop disease. If we populate these animals with a complex bacterial flora an early onset, caecal-dominated disease develops. Selective colonization with *Enterococcus faecalis* leads to distal colonic disease. In contrast colonization with *Escherichia coli* produces caecal-dominated disease. Conversely in mice the response to a given bacterial colonization varies according to the genetic background. Thus *E. coli* produces gastrointestinal inflammation in the IL10 knockout mouse but not in the HLA-B27 transgenic mouse, whereas *Bacteroides vulgaris* produces disease in the HLA-B27 transgenic mouse but not in the IL10 knockout mouse.

Ghosh: One of the problems you illustrated is the fact that there are multiple factors impinging on the phenotype. Once a patient is diagnosed there are pharmacological and surgical measures that are difficult to control for. Have you tried looking at phenotype at presentation and first diagnosis as separate from phenotype overall?

Ahmad: We can do these analyses but one needs to consider that disease phenotype is not stable with time. A number of studies have now shown that both disease location and behaviour change with time. In our genotype–phenotype studies we attempted to minimize this problem by classifying phenotype in patients who had been under regular follow-up for many years, not just at diagnosis, but at multiple time points. I agree, however, that it would be interesting to investigate associations with phenotype at first presentation. We have a prospective cohort of UC patients recruited eight years ago which we will use to look at this. We chose UC because eight years ago we thought we were making more progress with UC rather than CD genetics.

Schölmerich: The model may be even more complicated. We took our bone marrow transplantation population where we have DNA from the donors and recipients. When we typed them for *CARD15* it turned out that when we had a donor and a recipient with a mutation all recipients died, from graft versus host disease and sepsis. If we took cases where either donor or recipient had a mutation, the mortality was about 40–60%. The overall mortality where neither recipient nor donor had mutations was less than 30%. It depends on whether you have one cell population affected by a genetic defect or two, and it means that the expression of the genetic abnormality must be taken into account if you consider the interaction with the environment, which in this case was bacteria.

Jewell: I am not sure I understand. Do you see an increasing effect as to whether the donor or recipient are simple or compound heterozygotes or homozygotes?

Schölmerich: If there was one mutation in either the donor or recipient, then mortality doubled. But if the donor and recipient both had a mutation, the mortality was 100%.

Jewell: You talked about *CARD15* expression in lymphocytes and epithelial cells, which are not standard cells to be expressing this.

Schölmerich: No, we looked for the genetic abnormalities in the DNA we had frozen not for expression in specific cells. When you do a bone marrow transplantation you kill the immune system in the recipient. Thus, from the donor you will have immune cells such as for example lymphocytes; in the recipient it will be mainly epithelial cells having a mutation.

Meddings: I was interested in the bone marrow transplant story. I thought there were recent data suggesting that post bone-marrow transplant, some epithelial cells seem to be derived from the donor rather than the recipient. Could this be your *CARD15* coming into an epithelial cell that was bone-marrow donor derived?

Schölmerich: It could be. I am not sure how many of these bone marrow cells have this chimerism. We didn't look for this; it would be an interesting aspect.

Ahmad: Adoptive transfer of inflammatory bowel disease by haematopoietic cells is recognized in several animal models of inflammatory bowel disease and remission of Crohn's disease has been reported in patients who have received a bone marrow transplant. Two recent case reports have suggested that adoptive transfer of Crohn's disease can occur with allogeneic stem cell transplant from donor to recipient. We recently reported a case of fulminant Crohn's colitis that occurred following non-myeloablative allogeneic stem cell transplantation for Hodgkin's lymphoma (Sonwalkar et al 2003). We speculated that adoptive transfer of Crohn's disease may have occurred and performed a genetic analysis on known susceptibility loci for significant donor–recipient mismatches. Although the genetic analysis was not particularly heavyweight we noted that the donor and recipient had several haplotype mismatches in HLA class III genes at the *IBD3* locus. In addition, the donor (but not the recipient) had a polymorphism of the 5′ UTR of *NOD2/CARD15* that may be associated with Crohn's disease. In response to this publication another group reported a similar case in which the recipient developed Crohn's disease after bone marrow transplantation. Importantly in this case pre-transplantation the donor had Crohn's disease and the recipient didn't. Genetic analyses were not carried out in this second report.

Cummings: I want to pull back a bit from the environment versus gene side. You showed data indicating that married couples who have very different phenotypes didn't share disease risk, which suggests that there isn't likely to be a strong environmental influence on the disease. But if you looked at their gut flora, you'd probably find they were hugely different, even though they lived together. The thing about the flora is that it is acquired at the beginning of life: you are born with a sterile gut and then acquire bacteria over two or three years. Many factors at birth or shortly after make a huge difference to your flora, including things like diet. For example, it has been shown that whether you excrete methane in your breath depends a great deal on whether your mother did. Environmental influences early in life may be very important. They may carry through into later life through things like the flora and the way that diet interacts with it. Just because you can't measure anything known in a married couple, or that they have very different phenotypes, doesn't necessarily mean that it is entirely genetically determined: there may have been environmental factors which operated earlier in life.

Rhodes: I was intrigued by your comment that *NOD2/CARD15* abnormalities were not associated with granulomas. If you take this in context with the fact that the discordant sib pairs have their phenotype determined very much by smoking, and presumably the ones with colonic Crohn's are likely to have a label of colonic Crohn's at least in part because they have granulomas, the inference from this would be that smoking may be an extremely strong factor determining the formation of granulomas. There is an enormously broad interaction between smoking and the immune system, which you can pick at to suit any hypothesis you like. But there is one paper (King et al 1988) showing that pulmonary macrophages from smokers are unable to kill phagocytosed *Listeria.* Have people looked for a correlation between smoking and granulomas in other types of Crohn's?

Jewell: I don't think they have. The question is how would you define whether someone has granulomatous or non granulomatous disease? You don't have the whole bowel to look at in sections all the way from top to bottom. There is one paper in the literature concerning a study I have unsuccessfully tried to interest several generations of pathologists in repeating, by Chambers & Morson (1979). They looked at the frequency of granulomas in jejeunal, right- and left-sided colonic disease. The proportion of granulomas went up as they progressively went down the gut. So in colonic disease (and particularly perianal) there was more than 80% granulomas. There have been some suggestions that smoking tends to go along with small-bowel Crohn's disease rather more strongly that with large bowel Crohn's. On these fragmentary bits of data, smoking would tend not to be stimulating granuloma.

Rhodes: I would be very surprised if that suggestion is correct. The colonic Crohn's disease phenotype seems almost to be determined by smoking status, certainly in mixed disease families (Bridger et al 2002).

Jewell: There are just too many variables, and we haven't designed the right study yet.

Ghosh: Most of the phenotypic definitions are by necessity made on the basis of morphology. But there are other ways of doing this, for example by raised C-reactive protein (CRP) versus non-raised CRP. In the cardiovascular literature there is a lot of interest in *CRP* gene polymorphisms.

Ahmad: We have just started to explore *CRP* polymorphisms with respect to infliximab response, but we haven't looked at it in terms of overall susceptibility or other phenotypes. The other way that people are now profiling Crohn's disease phenotype is by expression studies, looking at different patterns of gene expression.

Schölmerich: We should not overdo it. There might not even be such polymorphism in *CRP*. The problem is that in these studies the patients with elevated CRP respond to a drug while the others don't. This means that one has included the wrong patients. There are patients in these trials with chologenic diarrhoea or with a stricture which causes pain: both elevate the Crohn's Disease Activity Index (CDAI) which never really correlated with CRP in most studies. CRP measures inflammation. It is not surprising that an anti-inflammatory drug doesn't treat people who don't have inflammation. This is probably the cause of the strange findings that have now been reported from several trials.

Ghosh: CRP is a good quality control.

Schölmerich: That is what I was indicating. If more than 60% of patients don't have elevated CRP in such a trial, something is completely wrong in the trial design.

Kamm: You mentioned some early work suggesting that patients stay true to their disease in terms of phenotype. This has always been my clinical impression. However, Bill Sandborn has written recently that it is just a question of time: if you take a much longer view, most patients will develop extensive disease involving most parts of the gut. Perhaps at first presentation one can classify phenotype more carefully, but if you take a longer view perhaps it merges.

Ahmad: A number of studies have demonstrated that if you take patients with ileal disease, given long enough they will all develop stenoses. Similarly the distinction between stenosing and perforating disease is rather artificial as most small bowel fistulae will occur in the presence of stenoses.

Peña: We all agree that in a significant proportion of patients, ileal disease is confined to the ileum. If you leave it alone, they will develop strictures and then you remove the strictures. The puzzle is, if *NOD2* is related to ileal disease and to the development of strictures, why isn't it related to surgery? One possibility is that the people who did the study didn't follow the patients for a long time. I do not agree with the extreme view of Sandborn that if you wait 30 years, all patients will have extensive disease. The evidence does not support this.

Fedorak: If we take the Vienna classification of Crohn's disease, it has become clear that, over time, disease location changes in about 15% of cases, and disease behaviour changes in about 80%. Thus these phenotypic determinations of Crohn's disease are not stable over time. Nevertheless, it is critical that we develop some form of disease classification in order to better predict response to therapeutic agents. In anticipation of this there was great hope that a molecular and genetic classification would meet these objectives. Yet I am hearing from the preceding discussions that we are a long way from classifying IBD genetically.

Since a genetic classification is likely to be the only viable and stable classification, shouldn't we be putting some resources and energy into this?

Ahmad: We certainly believe that we should be directing our efforts towards this goal. The Vienna classification is the best we have at the moment. In time we believe that this will be complemented by a genetic classification. One of the major obstacles currently challenging genetic investigators is the accurate clinical characterization of these patients.

Fedorak: Is part of the problem with developing a molecular genetic classification that we have not adequately developed the corresponding phenotypic databases?

Ahmad: It is not only crucial to have a detailed phenotypic database but also to keep it up to date. This is extremely time-consuming.

Fedorak: Are there any worldwide efforts in putting together databases from large centres and keeping them accurate?

Ahmad: The IBD Genetics Consortium comprises investigators from 11 worldwide centres. The emphasis of this group for the last few years has been the pooling of linkage data with analyses by IBD type (CD, UC or IBD only). In the last 12 months they have begun to carry out analyses stratified by disease phenotype. There are clearly considerable difficulties with pooling data from multiple centres where the phenotype has been categorized by more than 30 different people. I think therefore that accurate phenotype studies will likely remain the domain of the single centre for the foreseeable future.

Rioux: The databases have to be what we call 'living' databases. Currently most are simply a snapshot in time. Setting up these databases to keep track over time of disease progression or change of diagnosis is very simple, compared with the real challenge, which is getting these data accurately in the first place. Molecular classification brings together environment and genetics. Potentially, in terms of prognostic tools, the strength of molecular classification will come from the combined examination of an individual's genetic variants (constant over time) and their current immune status (e.g. by examining a gene expression pattern of circulating immune cells).

Jewell: The problem with molecular classification is that it is going to be a molecular classification for different populations of people. This may have limited usefulness. For example, USA and most of Europe have similar *NOD2* mutation rates in Crohn's disease. But if you then go to Ireland, Scotland and Scandinavia there is a 50% reduction, and China, Korea and Japan don't use *NOD2* mutations at all in their Crohn's disease. Yet none of us who have spent any time in Japan think there is any clinical difference between Crohn's disease there and here. Japan is a particularly interesting example: their Crohn's patients don't have *NOD2* mutations and they don't have the *IBD5* haplotype. Molecular classification may happen, but it will need to be specified for different populations.

What we are beginning to see, however, is that we can predict how the disease is going to behave to a degree. For those who are going to have mild disease without extra-intestinal manifestions, this might be a useful thing to tell people at diagnosis. This raises ethical issues about whether insurance companies will be allowed access to these data.

Bjarnason: I want to ask about the bone marrow transplants. Once you have a bone marrow transplant, how difficult is it to distinguish between Crohn's disease and the effects of cytotoxic drugs, immune suppression, graft-versus-host disease (GvHD) and so on?

Ahmad: I think it is extremely difficult. In the case we reported histological features were not compatibile with acute or chronic GvHD, which are characterised by crypt epithelial cell apoptosis and segmental crypt loss, or fibrosis of the submucosa, with or without a dense lymphocytic infiltrate respectively. It may be difficult to distinguish Crohn's disease from an infective colitis particularly in immunosuppressed patients receiving cyclosporin. Herpes simplex virus (HSV) colitis in particular may be indistinguishable from Crohn's disease. Polymerase chain analysis for HSV DNA or immunohistochemistry may be helpful in identifying HSV colitis.

Bjarnason: I would be reluctant to push my research career along on the basis of that observation, because of the confounding effects of the diseases.

Stange: What happened to hygiene? There has been an English study clearly showing that hygiene seems to be important during childhood. One point that touches this that we haven't discussed at all is that there are populations in which Crohn's disease seems almost unheard of. They have very poor hygiene, parasites and so on.

Ahmad: It might be interesting to speculate how the hygiene hypothesis fits together with our current knowledge of the genetics of IBD.

Peña: The fascinating thing is that atopy, allergy and autoimmune disease coincide and are at a higher frequency in the northern populations. This is why hepatitis A is a very good marker for the hygiene hypothesis. If you looked for something in Europe 50 years ago, 100% of the population would have had antibodies against hepatitis A. But now in The Netherlands, for example, only 20% of the population born after the Second World War have these virus antibodies. This is why some investigators have recently used the virus as a good marker for the hygiene hypothesis (Termorshuizen et al 2000). Infectious organisms, bacteria in the lumen of the intestine and the hepatitis A virus, may normally induce the development of regulatory T cells and protective immunity that limits airway inflammation and promotes tolerance to respiratory allergens. In the absence of infections, Th2 responses may predominate (Umetsu et al 2002). The microflora may contribute to determine the different distribution in different countries. In Israel, there is no association of *CARD15* mutations with ileal disease or specific disease course or behaviour (Fidder et al 2003). In other Jewish populations, for example the Chuetas who live in Mallorca, Spain, suffer from Mediterranean fever. Several mutations in chromosome 16 have been associated with this disease (Domingo et al 2000).

These mutations are not in the same region as the ones found in the same chromosome in the *CARD15* gene. As far as we know, the Chuetas do not suffer from Crohn's. Perhaps what this is telling us is that the microenvironment (the gut flora, for example) is different in these two populations, even though they have a common pool of genes. Therefore, the point raised by Eduard Stange is a fascinating one, and perhaps the defensins and their genes that also vary in different populations will be giving us some clues about differences in disease rates between populations even when they share common ancestors.

Jewell: It is probably a lot more complicated than that. We have just got some data supporting an association between a polymorphism in the *MEFV* gene and Crohn's disease. It is possible that all these observations are just being confounded by linkage disequilibrium and the true association is with a neighbouring gene. I would have to come back to John Rioux and ask are haplotype blocks constant across populations, or do they vary within populations?

Rioux: What we see is that the structure is quite similar from one population to another. The differences will be the frequencies of patterns or haplotypes within a particular block. The major differences in actual block definitions will be, as I mentioned in the African-derived populations, that there is much greater diversity, so the blocks are smaller but have more haplotypes. We don't have a lot of data yet, but it may be that the larger blocks in Caucasians are simply a combination of a couple of smaller blocks that haven't been broken up in this younger population. Mostly the structure is quite similar and haplotypes or patterns are nearly identical across populations. The differences in frequencies could have an important impact.

Jewell: One interesting observation is that UC in the immigrant Asian population in the UK is as high, if not higher, than the prevalence in Caucasians. The genes haven't changed. We know UC is present in India, but at a 10-fold reduced frequency. This suggests powerfully that the environment is important.

Furrie: We can't overestimate the amount of immune insult a multicellular parasite will cause to the immune system. If you take mice that are susceptible to diabetes and they get a worm infection, they will never develop diabetes. They are too busy dealing with this multicellular parasite to initiate the autoimmunity. If you push the immune system down that route, it will not necessarily do what its genetics might determine.

Rioux: This comes back to John Cummings' comment. You have a genetic background. You are given a certain number of tools and it depends on what you do with these tools. Whether you are dealing with the worms now (or you dealt with them 20 years ago) has a huge impact on the memory of your immune system. Even though your immune system has a genetic background, your environment has a huge impact on what it will recognize or interact with at the present time.

Jewell: Can anyone bring us up to speed with the trial of round worms for treating IBD?

Sartor: I spoke recently to Joel Weinstock. He has been involved with a multicentre controlled trial of a porcine roundworm in treatment of IBD. He said there was a 50% improvement rate in the UC population. He has not yet broken the code in the Crohn's disease population. He didn't say what the placebo rate was, although he says it is statistically significant.

Schölmerich: I have seen two papers from this group. One of them was in a mixed population and one in Crohn's. Both showed significant differences between the placebo and worm groups. Both studies look convincing. Even if the phenotype is expressed in patients with early childhood hygiene, later on in life, we might be able to reset the system with this approach.

Meddings: Is it more important to understand the overall genetic background of the host, or is it more important to understand what genes are expressed at the tissue level? How would we go about doing this?

Rioux: It is impossible to say that either one is key. We can't look at one without looking at the other. Human nature, however, is to specialize in one rather than the other.

Sartor: There are many studies looking at microarrays in Crohn's and UC. The problem is that you may be at the end of the spectrum rather than at the beginning of the pathogenesis. In all the animal models, regardless of how one gets there, you are at the same destination with a similar immune response. The concept of the immune defect may not be how it initiates, but once initiated you do go down common pathways.

Ghosh: Going back to the *NOD2* story, the Blau syndrome where the mutation is not in the leucine rich-repeat but in a different area, has chronic inflammatory features but they are different from Crohn's disease. Secondly, some of the data on the inbred mouse model which Warren Strober reported in gastroenterology last year showed that if you have colitis-susceptible mice, chromosome 11 coding for p40 IL12 is implicated, but you can make them resistant to colitis if you block TLR4 (Bouma & Strober 2003). The mutations in the sensor and the effector arm are not necessarily in synchrony all the time, but may balance each other.

Rioux: It has been shown many times in simple Mendelian diseases that the same gene mutated in different ways can give completely different phenotypes.

Sartor: It is important to make note of the identical twin data, which emphasize the importance of the environment. There is a lack of concordance, particularly in UC, where it is 10% at best, and 44–50% in Crohn's disease.

Jewell: That is just highlighting the difference between whether we are describing disease susceptibility genes or phenotype determining genes. I don't think we have a disease susceptibility gene yet.

Sartor: I agree, but this clearly shows the importance of the environment on the initiation of a response. The question that I have never seen answered by the twin data—and this may be unanswerable because of the reasonably small size of that population—is what is the phenotypic concordance in identical twins regarding extra-intestinal manifestations, locations of the disease and fistulizing versus stenotic variants?

Ahmad: As far as I am aware phenotypic subgroup analyses have not been carried out on the twin pair data sets. Any such analyses would indeed be limited by sample size.

Peña: There was an editorial recently trying to clarify and understand what it means if you find a functional single nucleotide polymorphism (SNP) in a population that is associated with a disease, and you go to another population and the same functional SNP is not associated (Vercelli 2003). Dr Vercelli described and commented on the association of a *CD14* gene polymorphism in a population and the lack of association in others. The CD14 receptor together with TLR4 are important in the response to LPS, but the genes and environment (the presence of a determined intestinal microflora) determine in combination whether you are going to have either a Th1 or Th2 response. This tells us that the immune response is very complicated. Before we use microarrays and other sophisticated technology, we should do clustering of all the information we have in all the genes that we know that are important in the proinflammatory and regulatory response to Th1 and Th2. It is possible that we will start to see clustering that can be linked to the clinical disease. In other words, I think the reason why we can't reproduce the different findings of association studies in different populations is first because we are dealing with different populations with different microflora, and secondly because we are looking at just one or two suspects when we should be looking at all of them.

Rioux: I agree, but we have to remember that the alternative explanation for that observation is that there is not replication. This is one thing we have to worry about when we see differences between studies, but you are right that this is not always obvious.

Rhodes: Francis Crick, in his book *The amazing hypothesis*, tells an anecdote of some chap attending some congress of firemen in the states. His buddies come up to him and tell him of another participant who is his spitting image. It turns out that it is his identical twin from whom he'd been separated at birth. Not only has he grown the same moustache and has become a fireman, but he has the same bizarre hobbies. This is illustrating the influence of genes on environment, which is difficult to unpick. I have a question: should pANCA and ASCA be included in this phenotyping? You mentioned the association with the same HLA subclass with colonic Crohn's and UC. Equally, the literature suggests the same association with pANCA and lack of association with ASCA. Does that indicate that isolated colonic Crohn's and UC are a spectrum of autoimmune diseases with phenotypic variations determined by smoking?

Jewell: I don't think we'll get into a debate about whether they represent autoimmune disease. I would take the stance that Crohn's isn't.

Rhodes: I'm specifically talking about isolated colonic Crohn's and UC, both of which have the same pANCA association.

Jewell: You could say for pure colonic Crohn's disease, that this is associated with pANCA rather than the ASCA, which has the same HLA associations that UC has, and there isn't an increased frequency of *NOD2* mutations. There may well be a case for saying that the genetic and serological background of colonic Crohn's disease is much more like UC. But I don't know where this gets you.

Rhodes: If pANCA, for example, correlated, as it does, with isolated colonic Crohn's and UC but didn't correlate with rectal Crohn's, then it might imply this is a separate phenotype related to some defect in handling bacteria.

Schölmerich: Before you include these laboratory measurements in any classification, you must be sure you are measuring the same thing. If you use four assays from different companies you'll likely get four different results.

Gibson: It is as precise as clinical assessment of phenotype!

Peña: There is some recent work headed by Dr David B. Sachar from the Mount Sinai in New York that has produced evidence that patients with perianal Crohn's disease represent an independent subgroup of the Vienna classification.

References

Bouma G, Strober W 2003 The immunological and genetic basis of inflammatory bowel disease. Nat Rev Immunol 3:521–533

Bridger S, Lee JC, Bjarnason I, Jones JE, Macpherson AJ 2002 In siblings with similar genetic susceptibility for inflammatory bowel disease, smokers tend to develop Crohn's disease and non-smokers develop ulcerative colitis. Gut 51:21–25

Chambers TJ, Morson BC 1979 The granuloma in Crohn's disease. Gut 20:269–274

Domingo C, Touitou I, Bayou A et al 2000 Familial Mediterranean fever in the 'Chuetas' of Mallorca: a question of Jewish origin or genetic heterogeneity. Eur J Hum Genet 8:242–246

Fidder HH, Olschwang S, Avidan B et al 2003 Association between mutations in the CARD15 (NOD2) gene and Crohn's disease in Israeli Jewish patients. Am J Med Genet 121A:240–244

King TE Jr, Savici D, Campbell PA 1988 Phagocytosis and killing of Listeria monocytogenes by alveolar macrophages: smokers versus nonsmokers. J Infect Dis 158:1309–1316

Sonwalkar SA, James RM, Ahmad T et al 2003 Fulminant Crohn's colitis after allogeneic stem cell transplantation. Gut 52:1518–1521

Termorshuizen F, Dorigo-Zetsma JW, de Melker HE, van den Hof S, Conyn-Van Spaendonck MA 2000 The prevalence of antibodies to hepatitis A virus and its determinants in The Netherlands: a population-based survey. Epidemiol Infect 124:459–466

Umetsu DT, McIntire JJ, Akbari O, Macaubas C, DeKruyff RH 2002 Asthma: an epidemic of dysregulated immunity. Nat Immunol 3:715–720

Vercelli D 2003 Learning from discrepancies: CD14 polymorphisms, atopy and the endotoxin switch. Clin Exp Allergy 33:153–155

Pharmacogenetics of inflammatory bowel disease

Dermot Kelleher, Richard Farrell* and Ross McManus

*Department of Clinical Medicine and Dublin Molecular Medicine Centre, Trinity College Dublin, Ireland and *Beth Israel Deaconess Medical Centre, Boston MA and Harvard Medical School, USA*

Abstract. Therapeutic outcome in inflammatory bowel disease has traditionally been related to disease activity. Recent data suggest that individual differences in drug disposition and metabolism, some of which are genetically determined, may play a significant role in the outcome of therapy for inflammatory diseases. Polymorphisms in the thiopurine methyl transferase gene (*TPMT*) are known to influence the outcome of therapy with azathioprine although pharmacogenetic analysis of outcome has not entered routine clinical use. The outcome of therapy with drugs such as steroids may be influenced by a wide range of genetic factors including polymorphisms in the multi-drug resistance 1 gene (*MDR1*), polymorphisms in glucocorticoid receptor genes and potentially other as yet undefined polymorphisms regulating the inflammatory process. Multiple polymorphisms have been identified in *MDR1* but their direct contribution to changes in expression and function have not as yet been defined. Lastly, as novel biological agents such as infliximab become established in clinical practice, it is clear that their therapeutic efficacy will likely be modified by polymorphisms downstream of their target molecules. New diagnostic and therapeutic algorithms are needed to directly determine the functional importance of the influence of host genetic factors on choice and dosage scheduling of therapy.

2004 Inflammatory bowel disease — crossroads between microbes, epithelium and immune systems. Wiley, Chichester (Novartis Foundation Symposium 263) p41–56

The human response to pharmacological agents is characterized by considerable heterogeneity, both in terms of efficacy and toxicity. Recent years have seen a considerable expansion of our understanding of the mechanisms underlying such heterogeneity. Polymorphisms in metabolic enzymes have been identified as contributing significantly to the development of adverse events in patients treated with commonly used agents. In addition, genetic heterogeneity in relation to receptors, targets and other molecules may also play a significant role in determining the outcome of therapy. In the case of chronic inflammatory bowel disease (IBD), therapeutic options have changed relatively little until recently. The treatment mainstays continue to be the use of 5-amino-salicylic acid derivatives, steroids and azathioprine/6-mercaptopurine with methotrexate utilized in certain patients who are refractory to therapy. In addition, treatment with cyclosporin has been utilized as rescue therapy in patients with aggressive ulcerative colitis. More recently, there has been considerable excitement generated by the use of the anti-TNF molecule, infliximab, in the treatment of refractory Crohn's disease. With regard to the common agents used, the drug metabolic profile of azathioprine provides for an important and well-established model of pharmacogenomic responses to therapy which illustrate many of the advantages and drawbacks in the translation of pharmacogenomics to clinical practice. However, as illustrated, pharmacogenetics may play a significant role in the metabolism and disposition of many commonly used drugs.

TPMT genes and azathioprine

The enzyme thiopurine-methyl-transferase (TPMT) catalyses the conversion of 6-mercapto-purine, the active ingredient of azathioprine, to the catabolic metabolite 6-methyl-mercapto-purine (6-MMP). Deficiencies in TPMT activity result in the accumulation of the 6-thio-guanine (6-TG) nucleotides, which are immunosuppressive and myelo-toxic. Hence, polymorphisms in the *TPMT* gene have the potential to significantly influence the outcome of therapy (Weinshilboum 2001). The identification of genetic heterogeneity in *TPMT* resulted in a classification of individuals based on polymorphisms within this gene. Individuals who are homozygous for alleles with normal activity account for approximately 89% of a normal population and would have, in general, normal to high activity of the enzyme expressed. Individuals who are heterozygous have intermediate activity and account for 11% of the population. Approximately one in three hundred individuals in the Caucasian population are homozygous for low-expressing forms of the *TPMT* genes and hence have low-to-absent *TPMT* activity. Critically, such individuals are at high risk for the development of severe myelo-toxicity, which is dose-related in response to azathioprine therapy. Heterozygotes may also be at slightly increased risk of myelo-toxicity and may benefit from dose reduction.

By contrast, individuals who are high expressers have high levels of 6-methyl-mercapto-purine, which is associated with hepato-toxicity in inflammatory bowel disease. Such individuals may also fail to respond to therapy at conventional doses. Hence, it would seem reasonable that individuals with chronic inflammatory bowel disease, scheduled for therapy, should be tested for *TPMT* genotype prior to commencement of treatment with azathioprine. However, this point of view is not universally held as pharmacogenetic analysis may identify only a subset of individuals with expected side-effects from therapy. Bone marrow toxicity in patients with azathioprine is not infrequent. In a group of 41 patients with leukopaenia or thrombocytopaenia, reported by Colombel et al (2000), 73% of patients had wild-type activity and only 10% were TPMT deficient. Therefore, it remains controversial as to whether *TPMT* genetic testing should be mandatory for patients with inflammatory bowel disease or whether careful monitoring with both blood counts and liver function tests will identify the majority of patients who will develop toxic events. However, it does appear apparent that clear identification of individuals with a genetic tendency to develop toxic responses to azathioprine will permit dose modification to appropriate levels in those patients. Nonetheless, at this point in time dosage schedules for such modifications have not been formalized.

Steroid therapy in inflammatory bowel disease

Glucocorticoids remain the mainstay of treatment of active chronic IBD. However, in patients with both ulcerative colitis and Crohn's disease, a significant proportion of patients fail to respond. Many other patients become steroid-dependent and are unable to withdraw from corticosteroid treatment either as a result of suboptimal initial therapy, or as a result of high levels of disease activity. Corticosteroids mediate their anti-inflammatory responses by binding the intra-cellular glucocorticoid receptor, also known as the classic glucocorticoid receptor, GR or GRα that is transported to the nucleus of the target cell rapidly with immediate effects on transcriptional regulation. GR has significant effects on the transcription of genes such as NF-κB and AP1, which are potent transcription factors for pro-inflammatory cytokines and adhesion molecules. In addition, corticosteroids induce the inhibitor I-κBα, which binds and sequesters NF-κB in the cytoplasm (Deroo & Archer 2001). A significant proportion of patients are resistant to the effects of corticosteroids although exact numbers may vary due to the lack of a clear universally held definition in this condition. Recent studies have demonstrated that patients who have steroid resistant disease also manifest steroid resistant lymphocytes in that lymphocyte proliferation is inhibited by less than 60%, even at super-physiological glucocorticoid concentrations. (Hearing et al 1999). Research in impaired sensitivity to glucocorticoid inhibition in IBD has identified three potential molecular mechanisms:

- Decreased cytoplasmic glucocorticoid concentration, as a result of increased drug efflux, mediated through the P glycoprotein 170 molecule coded for by the multi-drug resistance (*MDR1*) gene.
- Impaired glucocorticoid signalling because of dysfunction at the glucocorticoid receptor.

- Constitutive disease activity, including for example the activation of epithelial pro-inflammatory mediators, such as NF-κB.

The *MDR1* gene encodes the P glycoprotein 170 (Pgp170) molecule, which is expressed in both lymphocytes and gastrointestinal epithelial cells. In previously published studies, we have demonstrated that high levels of Pgp170 expression on peripheral blood lymphocytes correlates with failure of therapy in both ulcerative colitis and Crohn's disease. The *Mdr1* knockout mouse also demonstrates the importance of this molecule for steroid efficacy. In this model there is at least a twofold increase in colonic steroid levels in mice treated with dexamethasone (Schinkel et al 1995). Of note, more recently, it has been demonstrated that the *Mdr1* knockout mice develop IBD, possibly as a result of failure of clearance of bacterial toxins. In addition, *Mdr* gene polymorphisms have been implicated in the pathogenesis of ulcerative colitis (Schwab et al 2003a). High Pgp170 expression has also been demonstrated in patients with resistant rheumatoid arthritis, rejecting renal graft recipients and in patients with drug resistant epilepsy (Ho et al 2003).

In a previous publication, we have demonstrated that high expression of Pgp170 was associated with resistance to steroid therapy in patients with both ulcerative colitis and Crohn's disease (Farrell et al 2000). High expression of Pgp170 was found in peripheral blood lymphocytes of patients with ulcerative colitis who had undergone colectomy for disease and this high level of expression was persistent even several years post colectomy. Similarly patients with ulcerative colitis who had resections for refractory disease also exhibited high levels of Pgp170. There was a significant correlation between levels of Pgp170 on circulating and intestinal lymphocytes. In view of this stable high expression of Pgp170 in refractory patients we have postulated a genetic cause for steroid refractory disease.

Polymorphisms of the *MDR* gene, and consequences for gene expression and protein activity

The first in-depth analyses of mutation of the *MDR1* gene came about as a consequence of its being somatically mutated in cancers. In recent years the focus has shifted to constitutionally inherited polymorphisms that could alter the activity or expression of the gene. To date, more than 25 polymorphisms in the *MDR1* gene have been described (Table 1), however the majority of these are non-coding and thus do not alter P glycoprotein sequence (for recent reviews see Fromm 2002, Schwab et al 2003b, Sakaeda et al 2002).

Polymorphisms causing non-synonymous changes to the coding regions are prime candidates for alteration of the function of the protein. The G2677T/A

TABLE 1 Polymorphisms of *MDR1* affecting the coding sequence

Exon	*Allele*	*Allele frequency*	*Reference*
Exon 2:	A61G–Asn21Asp	88.8/11.2	(Cascorbi et al 2001)
Exon 5:	T307C–Phe103Leu	99.5/0.5	(Hoffmeyer et al 2000)
Exon 7:	A548G–Asn183Ser	98.6/1.4	(Kim et al 2001)
Exon 11:	G1199A–Ser400Asn	94.5/5.5	(Cascorbi et al 2001)
Exon 13:	C1474T–Arg492Cys	98.6/1.4	(Kim et al 2001)
Exon 21:	G2677T/A–Ala893Ser/Thr	56.5/41.6/1.9	(Cascorbi et al 2001)
Exon 24:	G2995A–Ala999Thr	94.5/5.5	(Mickley et al 1998)
Exon 26:	A3320C–Gln1107Pro	99.8/0.2	(Cascorbi et al 2001)
Exon 26:	T3421A–Ser1141Thr	99.7/0.3	(Fromm 2002)

Note: Allele frequencies are quoted from the largest available studies. For the sake of consistency, frequencies are given for caucasian populations in this table.

polymorphism is the most common variant of the *MDR1* gene and causes a residue change from an alanine to a serine and more rarely to a threonine at position 893 (Mickley et al 1998, Cascorbi et al 2001). P-glycoprotein molecules with the substituted serine were shown to be more active, leading to enhanced efflux of digoxin from cells (Kim et al 2001) and were associated with altered drug efflux characteristics in the MCF-7 cancer cell line (Fairchild et al 1990). However, these findings were not replicated in a study in which the cell surface expression and function of Asn21Asp, Phe103Leu, Ser400Asn, Ala893Ser and Ala999Thr *MDR1* variants were tested and found not to differ from wild-type (Kimchi-Sarfaty et al 2002).

A polymorphism that does not affect the coding sequence but has been associated with differences in the expression of the *MDR1* gene is C3435T, located in exon 26. This is a silent polymorphism affecting a tRNA wobble position and was reported by Hoffmeyer et al (2000) to have a frequency of 46.1/53.9 in a German population. In intestinal tissue, it was shown that CC homozygotes for the C3435T polymorphism had an almost twofold increased expression of P-glycoprotein as compared to TT homozygotes, with heterozygote expression at an intermediate level. This trend was also seen in a Japanese study of human placental tissue, although it did not reach statistical significance (Tanabe et al 2001). This group also showed an association for alleles at the 2677 site, with P-glycoprotein expression in the placenta, which again was not significant. However, they did find a significant association of wild-type (TT) homozygotes of the −129 *MDR1* promoter polymorphism with higher expression of P-glycoprotein, compared with TC heterozygotes.

A study of lymphocytes derived from a German population sample showed a higher expression of *MDR1* mRNA in cells homozygous for the 3435 CC allele with lower levels in CT and TT carriers respectively ($n=31$) (Hitzl et al 2001). However, a Japanese study found the opposite: TT homozygotes were shown to be higher expressers of the *MDR1* gene than persons carrying the C allele either alone or in combination ($n=13$) (Nakamura et al 2002). In both studies efflux activity of the protein mirrored the findings of variable gene expression. In $CD56^+$ T cells there were significant differences in P-glycoprotein activity as determined by rhodamine efflux assays (Hitzl et al 2001). This finding was replicated in another European study of 59 HIV patients where CC homozygotes were shown to have a higher expression of mRNA and protein compared to either other genotype (Fellay et al 2002).

Thus it would appear that there is an association with both protein and mRNA expression and activity for the 3435 polymorphic site. The results are somewhat contradictory however, with European and Japanese cohorts showing different associations. Since the 3435 site does not alter the protein sequence, it would appear most likely that the effects on gene expression associated with this SNP may well be due to other SNPs which are physically linked to it. Therefore the 3435 may be strongly associated with a polymorphism at a different site, which either affects MDR expression or function. Since patterns of association may differ between populations it is possible that the variable effects on gene expression seen in different groups may be due to altered patterns of SNPs in haplotypes. As described in the next section, polymorphism frequencies are known to differ on a regional basis.

Polymorphisms and differences in frequency between populations

The frequency of polymorphisms of the *MDR* gene has been assayed in several populations, particularly for the exon 26 C3435T site (reviewed by Schwab et al 2003b). In general, the frequencies of both alleles fall within the 0.4–0.6 range in European populations that have been tested; that is they are almost equally prevalent (Hoffmeyer et al 2000, Kim et al 2001, Cascorbi et al 2001, Schaeffeler et al 2001, Ameyaw et al 2001). In the majority of reports the C allele is slightly less prevalent (0.43–0.54) than T. A number of Asian populations have also been sampled; here the frequency range is similar, but in these populations the C allele (0.53–0.61) is more common than T (Tanabe et al 2001, Schaeffeler et al 2001, Ameyaw et al 2001, Sakaeda et al 2001). The sole exception to this trend is a south west Asian population where the C allele is relatively rare (0.34) (Ameyaw et al 2001).

In stark contrast to the relative equal frequencies of alleles at site 3435 in Europeans and Asians, there is a much higher preponderance of the C allele in

Africans that have been tested (Schaeffeler et al 2001, Ameyaw et al 2001). In these groups, the C allele has a frequency ranging from 0.73 up to 0.90. This results, in practice, in a very small proportion of the population being carriers of the TT genotype—reaching a maximum of 6% of individuals in a Sudanese sample. This represents a striking regional disparity although the causes are not clear. One obvious possible interpretation is that homozygosity for the T allele represents a selective disadvantage for the host.

Reports on the frequencies of other *MDR1* alleles are patchy. The 2677T allele appears to be present in a German and Japanese population at similar frequencies (0.42), but as for 3435T, 2677T is markedly less common in Africa (0.13).

The functional relevance of these findings is open to speculation. It has been noted that the CC genotype of the 3435 site is associated with the highest expression of *MDR1* in a European Caucasian group and if this functional linkage is true also for African populations, this could indicate that high activity of MDR is strongly selected in Africans. Since P-glycoprotein is important in protecting against harmful substances of various provenance, a rational case for high expression can be made in areas where infections and toxic agents in the diet may occur. However, why such effects should not also be felt in Asia and Europe is more of a mystery given that similar considerations would surely have prevailed in these regions until recently.

As previously stated, more than 25 *MDR1* polymorphisms have been identified and a polymorphism in exon 26 (C3435T) has been shown to correlate with levels of expression in healthy individuals. Homozygotes for this polymorphism have high levels of *MDR1* expression and function. A number of studies have reported on a possible association of the *MDR1* gene and inflammatory bowel disease. The C3435T polymorphism has been demonstrated to be increased in one German population study with chronic ulcerative colitis (Schwab et al 2003a) but not another (Croucher et al 2003). Neither did Brant et al (2003) replicate this finding in a large IBD cohort, however they did find an association between the 2677 G allele and IBD. While these findings might be seen as inconsistent, it is clear that the Pgp170 molecule fulfils many functions in the gut including potentially clearance of xenobiotics in the normal gut and therapeutic agents in disease.

Steroid resistance and glucocorticoid receptor abnormality

The further complexity of pharmacogenomics of steroid therapy is illustrated by the involvement of inherited abnormalities of glucocorticoid receptors. Increased glucocorticoid receptor β (GRβ)-specific messenger RNA has been reported in a Japanese group of patients with steroid-resistant ulcerative colitis (Honda et al 2000). These findings mirror similar studies in steroid-resistant asthma and

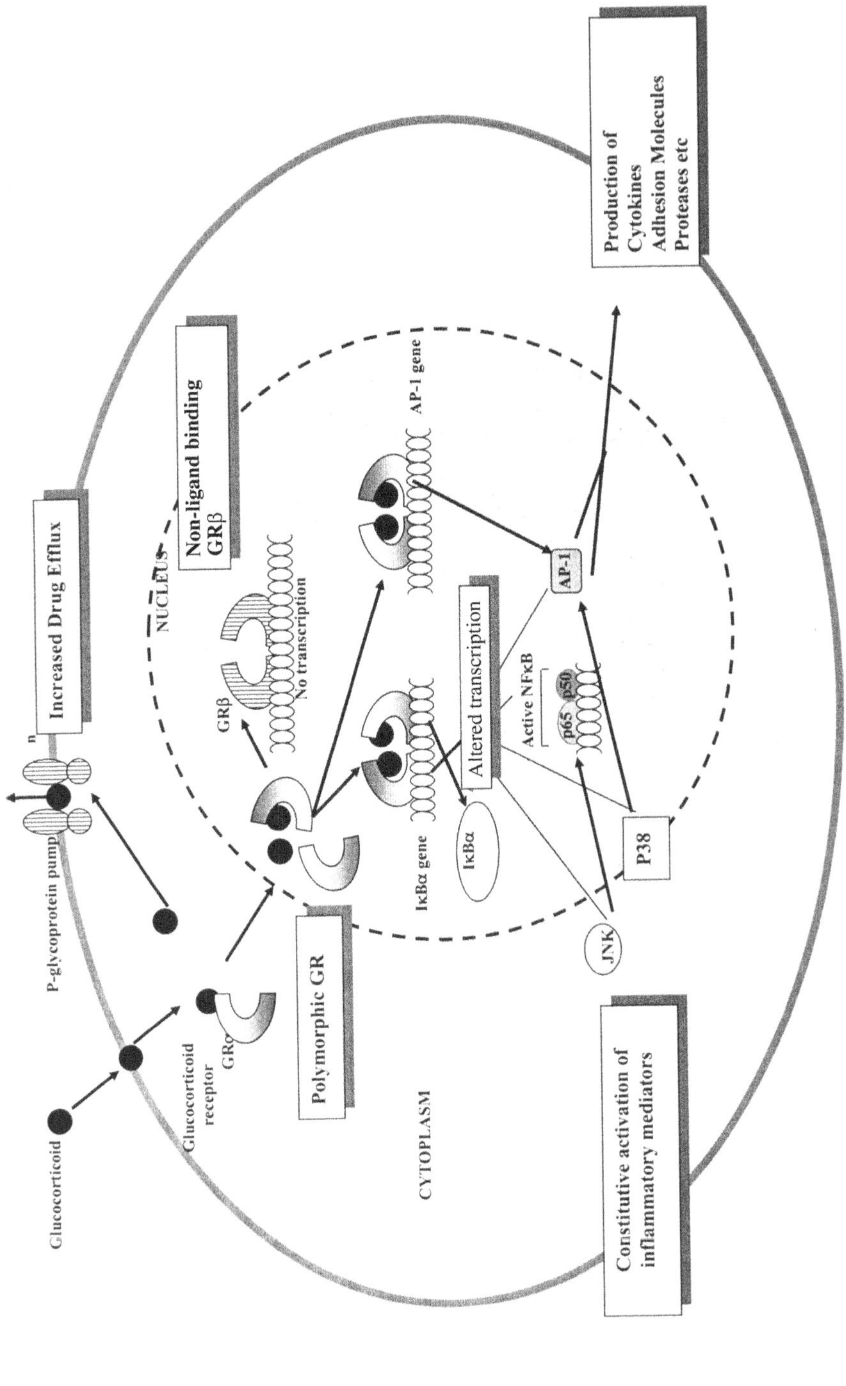
Glucocorticoid
Glucocorticoid receptor
GRα
P-glycoprotein pump
Increased Drug Efflux
Non-ligand binding GRβ
NUCLEUS
GRβ
No transcription
AP-1 gene
Polymorphic GR
IκBα gene
IκBα
Altered transcription
Active NFκB
p65
p50
AP-1
P38
JNK
CYTOPLASM
Production of
Cytokines
Adhesion Molecules
Proteases etc
Constitutive activation of
inflammatory mediators

FIG. 1. Mechanisms of glucocorticoid resistance in inflammatory bowel disease. Mechanisms identified or suggested as altering steroid sensitivity in inflammatory bowel disease include the following (i) reduced cytoplasmic glucocorticoid concentration secondary to increased P-glycoprotein mediated steroid efflux resulting from increased activity or overexpression of the multidrug resistance gene (*MDR1*); (ii) altered expression of glucocorticoid receptor β (GRβ), a truncated splice variant of the normal isoform GRα, that does not bind glucocorticoid ligands, and is therefore unable to transactivate glucocorticoid-responsive genes; and (iii) modulation at a functional level resulting from constitutive activation of proinflammatory mediators, NF-κB, AP1 and upstream protein kinases p38 and c-Jun N-terminal kinase (JNK) which can directly inhibit (thin arrows) the anti-inflammatory action of a limited number of GRα molecules by preventing GR transcriptional activity or by altering the threshold for anti-inflammatory action.

rheumatoid arthritis. However, such findings have not been universal in other groups and the levels of GRβ found in these studies have been relatively small. Hence, it is possible that these abnormalities represent a secondary effect, particularly as GRβ may be regulated by both cytokines and steroid administration.

Complexity in terms of steroid responses

It is very clear that the responses to glucocorticoids involve a complex set of interactions. Pharmacogenomic effects could be predicted to occur at many different sites in this sequence of interactions (see Fig. 1). Indeed, the effects of steroids in the regulation of I-κB transcription and hence the release of pro-inflammatory molecules such as tumour necrosis factor α may be subject to pharmacogenomic modification through interactions with a large number of potential activation molecules. It is notable that mutations in the *NOD2* gene responsible for a subset of familial Crohn's disease (Ogura et al 2001) may regulate the activity of NF-κB. Hence, the interactions of corticosteroids with transcriptional regulation depend on multiple genetic factors.

Pharmacogenomic response to anti-TNF therapy in patients with inflammatory bowel disease

Tumour necrosis factor (TNF) appears to play a key role as a pro-inflammatory cytokine in inflammatory bowel disease and in recent years the utilization of infliximab or humanized monoclonal anti-TNF antibody has revolutionized the treatment of severe refractory Crohn's disease. In studies on rheumatoid arthritis, patients with the *TNFA* 308GG genotype were found to be better infliximab responders than patients with AA or AG genotypes (Mugnier et al 2003). In patients with Crohn's disease, homozygotes for the *TNFA*-linked *LTA*1-1-1-1 haplotype showed an inferior response to infliximab and more recently it has been demonstrated that patients with polymorphisms in exon 6 of the TNF receptor 2 (*TNFR2*) also show a poor response (Taylor et al 2001).

Conclusions

Treatment of patients with inflammatory bowel disease is associated with very significant pharmacogenomic interactions. The TPMT pathway of metabolism of azathioprine-6-MP has long been recognized as an exemplar of the role of pharmacogenomics in the management of disease. Nonetheless, significant questions remain. Continuous monitoring of full blood counts and liver function tests in the early stages of therapy are mandatory, even in patients in whom the genotype is known. No dosing regimens have been formalized for the treatment of either high or low expressers. However, as therapy for inflammatory bowel disease enters the 21st century, it is possible that paradigms for treatment may shift. Currently, treatment is based on the induction of remission in individuals with high levels of disease activity. With further developments in genomics and potential genetic characterization of individuals at high risk of inflammatory bowel disease, it is likely that this paradigm will shift to one in which prospective patients may be treated based on predicted disease development. In such situations, the level of tolerable risk will be reduced and the utilization of pharmacogenomics to predict such risks would be mandatory.

Therapy with steroids remains a relatively unexplored area in terms of pharmacogenomics. Recent research indicates that the *MDR* gene coding for the Pgp170 molecule may be a key factor in regulating serum or cellular levels of corticosteroids. The recently identified polymorphisms within this gene appear to predispose to inflammatory bowel disease as well as associating with a relatively poor response to corticosteroids in subsets of these patients. Nonetheless, full characterization of the pharmacogenomics of corticosteroid therapy remains extremely complex. Future developments in this area will likely include the development of broad-spectrum pharmacogenomic chips that would permit the development of specific algorithms of therapy for patients with complex disease.

Lastly, pharmacogenomics is likely to play a role in the management of patients with novel biological therapies. Although *TNFA* and *TNF receptor* gene polymorphisms have been suggested to play a role in this area, it is very likely that the pattern of pharmacogenomics for this form of therapy will be increasingly complex. Genes regulating cytokine production such as p38 Map kinase, Map kinase kinases and phosphatases may all play a role in determining the eventual outcome in terms of cytokine secretion. Hence, further research involving complex genomics will be required before we can develop a clear picture of the nature of the response to such agents in the future.

References

Ameyaw MM, Regateiro F, Li T et al 2001 MDR1 pharmacogenetics: frequency of the C3435T mutation in exon 26 is significantly influenced by ethnicity. Pharmacogenetics 11: 217–221

Brant SR, Panhuysen CIM, Nicolae D et al 2003 MDR1 Ala893 polymorphism is associated with inflammatory bowel disease. Am J Hum Genet 73:1282–1292

Cascorbi I, Gerloff T, Johne A 2001 Frequency of single nucleotide polymorphisms in the P-glycoprotein drug transporter MDR1 gene in white subjects. Clin Pharmacol Ther 69: 169–174

Colombel JF, Ferrari N, Debuysere H 2000 Genotypic analysis of thiopurine S-methyltransferase in patients with Crohn's disease and severe myelosuppression during azathioprine therapy. Gastroenterology 118:1025–1030

Croucher PJ, Mascheretti S, Foelsch UR et al 2003 Lack of association between the C3435T MDR1 gene polymorphism and inflammatory bowel disease in two independent Northern European populations. Gastroenterology 125:1919–1920

Deroo BJ, Archer TK 2001 Glucocorticoid receptor activation of the I kappa B alpha promoter within chromatin. Mol Biol Cell 12:3365–3374

Fairchild CR, Moscow JA, O'Brien EE, Cowan KH 1990 Multidrug resistance in cells transfected with human genes encoding a variant P-glycoprotein and glutathione S-transferase-pi. Mol Pharmacol 37:801–809

Farrell RJ, Murphy A, Long A et al 2000 High multidrug resistance (P-glycoprotein 170) expression in inflammatory bowel disease patients who fail medical therapy. Gastroenterology 118:279–288

Fellay J, Marzolini C, Meaden ER et al 2002 Swiss HIV Cohort Study. Response to antiretroviral treatment in HIV-1-infected individuals with allelic variants of the multidrug resistance transporter 1: a pharmacogenetics study. Lancet 359:30–36

Fromm MF 2002 The influence of MDR1 polymorphisms on P-glycoprotein expression and function in humans. Adv Drug Deliv Rev 54:1295–1310

Hearing SD, Norman M, Probert CS, Haslam N, Dayan CM 1999 Predicting therapeutic outcome in severe ulcerative colitis by measuring in vitro steroid sensitivity of proliferating peripheral blood lymphocytes. Gut 45:382–388

Hitzl M, Drescher S, van der Kuip H et al 2001 The C3435T mutation in the human MDR1 gene is associated with altered efflux of the P-glycoprotein substrate rhodamine 123 from CD56+ natural killer cells. Pharmacogenetics 11:293–298

Ho GT, Moodie FM, Satsangi J 2003 Multidrug resistance 1 gene (P-glycoprotein 170): an important determinant in gastrointestinal disease? Gut 52:759–766

Hoffmeyer S, Burk O, von Richter O et al 2000 Functional polymorphisms of the human multidrug-resistance gene: multiple sequence variations and correlation of one allele with P-glycoprotein expression and activity in vivo. Proc Natl Acad Sci USA 97:3473–3478

Honda M, Orii F, Ayabe T et al 2000 Expression of glucocorticoid receptor in lymphocytes of patients with glucocorticoid-resistant ulcerative colitis. Gastroenterology 118:859–866

Kim RB, Leake BF, Choo EF et al 2001 Identification of functionally variant MDR1 alleles among European Americans and African Americans. Clin Pharmacol Ther 70:189–199

Kimchi-Sarfaty C, Gribar JJ, Gottesman MM 2002 Functional characterization of coding polymorphisms in the human MDR1 gene using a vaccinia virus expression system. Mol Pharmacol 62:1–6

Mickley LA, Lee JS, Weng Z et al 1998 Genetic polymorphism in MDR-1: a tool for examining allelic expression in normal cells, unselected and drug-selected cell lines, and human tumors. Blood 91:1749–1956

Mugnier B, Balandraud N, Darque A, Roudier C, Roudier J, Reviron D 2003 Polymorphism at position -308 of the tumor necrosis factor alpha gene influences outcome of infliximab therapy in rheumatoid arthritis. Arthritis Rheum 48:1849–1852

Nakamura T, Sakaeda T, Horinouchi M et al 2002 Effect of the mutation (C3435T) at exon 26 of the MDR1 gene on expression level of MDR1 messenger ribonucleic acid in duodenal enterocytes of healthy Japanese subjects. Clin Pharmacol Ther 71:297–303

Ogura Y, Bonen DK, Inohara N et al 2001 A frameshift mutation in NOD2 associated with susceptibility to Crohn's disease. Nature 411:603–606

Sakaeda T, Nakamura T, Horinouchi M et al 2001 MDR1 genotype-related pharmacokinetics of digoxin after single oral administration in healthy Japanese subjects. Pharm Res 18:1400–1404

Sakaeda T, Nakamura T, Okumura K 2002 MDR1 genotype-related pharmacokinetics and pharmacodynamics. Biol Pharm Bull 25:1391–1400

Schaeffeler E, Eichelbaum M, Brinkmann U et al 2001 Frequency of C3435T polymorphism of MDR1 gene in African people. Lancet 358:383–384

Schinkel AH, Wagenaar E, van Deemter L, Mol CA, Borst P 1995 Absence of the mdr1a P-Glycoprotein in mice affects tissue distribution and pharmacokinetics of dexamethasone, digoxin, and cyclosporin A. J Clin Invest 1995:1698–1705

Schwab M, Schaeffeler E, Marx C et al 2003a Association between the C3435T MDR1 gene polymorphism and susceptibility for ulcerative colitis. Gastroenterology 124:26–33
Schwab M, Eichelbaum M, Fromm MF 2003b Genetic polymorphisms of the human MDR1 drug transporter. Annu Rev Pharmacol Toxicol 43:285–307
Tanabe M, Ieiri I, Nagata N et al 2001 Expression of P-glycoprotein in human placenta: relation to genetic polymorphism of the multidrug resistance (MDR)-1 gene. J Pharmacol Exp Ther 297:1137–1143
Taylor KD, Plevy SE, Yang H et al 2001 ANCA pattern and LTA haplotype relationship to clinical responses to anti-TNF antibody treatment in Crohn's disease. Gastroenterology 120:1347–1355
Weinshilboum R 2001 Thiopurine pharmacogenetics: clinical and molecular studies of thiopurine methyltransferase. Drug Metab Dispos 29:601–605

DISCUSSION

Jewell: What does MDR1 do physiologically? Why have we got it?

Kelleher: It is a multifunctional pump that is there to get rid of toxic molecules. It is there to protect us. If you look at the gastrointestinal (GI) epithelium, *MDR1* is only expressed on the apical surface. This is probably why levels of the GI epithelium don't mirror those of the lymphocyte. Because it is specifically expressed on the apical surface it is perhaps delivering functionally at a higher level in the epithelium than it is in the lymphocyte.

Jewell: If it is important for mediating steroid resistance, why should a patient with ulcerative colitis (UC) who is proving resistant to intravenous steroids suddenly do well on cyclosporine?

Kelleher: Cyclosporine is both a substrate and an inhibitor of MDR. In general, if you want to use cyclosporine effectively in patients with severe refractory UC, you use a high dose. My theory is that this blocks the MDR1 pump and increases the intracellular level of corticosteroids. This can be demonstrated *in vitro*.

Jewell: It must be a saturable pump.

Kelleher: That is right.

Jewell: If you saturate with one drug, will it still pump out a totally different molecule? If you saturate it with steroids will it still pump out cyclosporine?

Kelleher: Some things seem to saturate and inhibit at the same time. Cyclosporine A can saturate but probably also has an inhibitory function. It is saturable and there are specific drugs which seem to have more of an inhibitory effect at a molar level than others.

Stange: There has been a study by D'Haens et al (2001) showing that corticosteroids are equivalent to cyclosporine given as monotherapy. This argues strongly against the notion that cyclosporine only acts by potentiating the corticosteroid effect. It probably acts on its own. We have the same experience with tacrolimus.

Kelleher: I wouldn't dispute the concept that cyclosporine has discrete effects, but in the setting of acute UC, you have to use high saturating levels to get a therapeutic effect. Similarly, cyclosporine is not terribly effective as a monotherapy in treating rheumatoid arthritis, but when it is added to methotrexate it greatly increases the potency.

Stange: The Dutch group also compared the two doses of 4 mg/kg and 2 mg/kg which had equivalent effects (Van Assche et al 2003). When we looked at the effect of oral tacrolimus a dose of 0.1–0.2 mg/kg appeared to be similar in effect to intravenous cyclosporine (Fellermann et al 2002).

Rhodes: I'd like to ask both Dermot Kelleher and Bill Roediger about the possible role of toxic phenols in all this. Bill, you had an interesting story a few years ago that UC was associated with a reduction in phenol sulfotransferase in the colon.

Roediger: Yes, there was a decreased capacity for phenol detoxification.

Rhodes: Since then, there have been at least two independent studies suggesting that paracetamol can be quite an important factor in inducing a relapse of UC, more so apparently than non-steroidal anti-inflammatory drugs (NSAIDs).

Kelleher: I haven't thought about that question. I don't know any specifics in terms of toxic phenols.

Rhodes: There are naturally occurring phenols that we are exposed to all the time and which are detoxified by sulfation.

Cummings: In the large bowel phenols come from the degradation of phenylalanine and tyrosine by bacteria, so it is part of an anaerobic process, which occurs in the absence of reasonable amounts of substrate.

Roediger: Most of the detoxification is actually an acetylation process in the cell. It is not a pumping process. There might be different levels at which detoxification works.

Parkos: I have heard two presentations talking about gender differences in these mice. The males get less severe disease than the females with regard to the IBD phenotype in these *Mdr*-deficient animals.

Kelleher: I don't know what potential mechanisms are involved in that. The development of disease is highly dependent on the bacterial flora. This seems to be a major determining factor. It is not like other models of colitis in that it doesn't have a major T cell component. It is more a permeability issue. T cell function seems to be similar in this model to that in the wild-type mice. It is complex.

Ghosh: Is there variability in the epithelial expression pattern of *MDR1* along the length of the gut?

Kelleher: Yes.

Ghosh: Is this up-regulated by inflammation?

Kelleher: The regulatory domain has an AP1 site and there is a part of this domain upstream that has an NF-κB site, hence it is potentially up-regulated by inflammation. The levels of expression differ at different levels in the GI tract, and also vary according to where you read about this: there are a number of publications that say it is predominantly expressed in ileum and less so in colon. There are other publications saying that colon has quite strong expression. It is very inducible. One of the studies done a few years ago used rifampicin (Greiner et al 1999), which produces a huge increase in the GI expression of *MDR1*. The figures in the paper don't show much expression in the epithelial lymphocytes. There may be some mechanisms by which overexpression of MDR is specific to the epithelial cells.

Gibson: If this is just a transport protein, you can surely saturate it. Yet the oncologists say that if you give more of the drug it doesn't work in people who have a high expression of *MDR1*, and clinical experience says that steroids don't work in higher doses. MDR1 is a very large molecule: it must be doing something else.

Kelleher: It does have functions other than pumping out drugs. Some years ago there was a publication in *Blood* which demonstrated a function for MDR1 in cytokine efflux (Drach et al 1996). This included efflux of both interferon and interleukin 4. In terms of the function, this seems to be heavily dependent on phosphorylation, particularly through the protein kinase C (PKC) pathway. To some extent, if you test this *in vitro* you will get very different levels of function depending on whether you expose lymphocytes (if you are using these as your model) to phorbol ester which activates PKC, or not. Though you might be saturating the resting molecule at lower levels, once it is phosphorylated the capacity to pump goes up dramatically. If you look at the situation in the gut in IBD, there is lymphocyte activation and an immediate step is activation of PKC. It could be far more related to functional activity.

Schölmerich: MDR1 is part of a family of ABC transporters, a group of some 70 different molecules expressed in the liver and all the outgrowings of the GI tract. Normally the GI tract expresses MDR1 to a lesser extent the further down you go. These 70 molecules taken all together transport everything. One of their main functions seems to be to transport lipids and toxic lipid products out of cells. They protect epithelial cells as well as the lymphocytes.

Gibson: I was wondering whether there was any non-transport function of MDR1.

Schölmerich: Not to my knowledge.

Meddings: Is anything known about up-regulation of this molecule not by things inside but by bacterial flora on the outside, or dietary effects?

Kelleher: Not a lot. St John's wort contains hypericin which up-regulates MDR1 and this may be ingested. In Sweden, it is made into an aquavit! *Helicobacter bilis* accelerates and *Helicobacter hepaticus* slows development of IBD in *Mdr*-deficient mice. This illustrates the difficulties people could have due to the complexity of the genetics of IBD itself. Simple changes in bacteria can modulate the development of colitis in an animal model like this. It makes it very difficult to look in the human gut and identify what changes in flora will be associated with IBD in a susceptible individual.

Gibson: We did a study a few years ago on rats, feeding them different oils. We found no evidence of direct inhibition of the Pgp efflux pump, using verapamil as the transported molecule (Vine et al 2002).

Ghosh: From what we know about thiopurine metabolism, can you explain why azathioprine is so slow in acting even if given intravenously?

Kelleher: That is a good question. One of the problems we have with molecules is that we don't understand how they work. We call azathioprine an immune suppressant but we aren't sure how it works in IBD.

Schölmerich: Markus Neurath's group recently published a paper in which they showed how it works (Thiede et al 2003). It blocks anti-apoptotic pathways. It takes a while to re-set this system so that apoptosis is no longer blocked. This is how he explained the delay in onset of effects.

Kelleher: In terms of apoptosis and anti-apoptosis this is probably where we will find the majority of IBD drugs actually do work including both azathioprine and infliximab. I raised the issue of whether we should be treating people who are well. It might end up being a safer form of therapy because you are treating fewer lymphocytes in their GI tract, so a lower level of apoptosis is needed.

References

D'Haens G, Lemmens L, Geboes K et al 2001 Intravenous cyclosporine versus intravenous corticosteroids: single therapy for severe attacks of ulcerative colitis. Gastroenterology 120:1323–1329

Drach J, Gsur A, Hamilton G et al 1996 Involvement of P-glycoprotein in the transmembrane transport of interleukin-2 (IL-2), IL-4, and interferon-gamma in normal human T lymphocytes. Blood 88:1747–1754

Fellermann K, Tanko Z, Herrlinger KR et al 2002 Response of refractory colitis to intravenous or oral tacrolimus (FK506). Inflamm Bowel Dis 8:317–312

Greiner B, Eichelbaum M, Fritz P et al 1999 The role of intestinal P-glycoprotein in the interaction of digoxin and rifampin. J Clin Invest 104:147–153

Thiede I, Fritz G, Strand S et al 2003 CD28-dependent Rac1 activation is the molecular target of azathioprine in primary human CD4+ T lymphocytes. J Clin Invest 111:1133–1145

Van Assche G, D'Haens G, Noman M et al 2003 Randomized, double-blind comparison of 4 mg/kg versus 2 mg/kg intravenous cyclosporine in severe ulcerative colitis. Gastroenterology 125:1025–1031

Vine DF, Charman SA, Gibson PR, Sinclair AJ, Porter CJH 2002 Effect of dietary fatty acids on the intestinal permeability of marker drug compounds in excised rat jejunum. J Pharm Pharmacol 54:809–819

Bacterial milieu and mucosal bacteria in ulcerative colitis

G. T. Macfarlane, E. Furrie[1] and S. Macfarlane

University of Dundee MRC Microbiology and Gut Biology Group, Ninewells Hospital Medical School, Dundee DD1 9SY, UK

Abstract. The aetiology of ulcerative colitis (UC) is unknown, but there is evidence that bacteria are needed for initiation and maintenance of the disease. A number of organisms have been associated with UC, but evidence for a specific transmissible agent is weak. Despite this, there is a good case for mucosal bacterial involvement, either through pathogens colonizing the epithelial surface, by non-pathogenic commensal species occupying adhesion sites on the mucosa and preventing invasion by harmful bacteria, or by inappropriate host immune responses to members of the normal microflora. Since mucosal bacteria exist in close juxtaposition to host tissues, it might be expected that they interact to a greater extent with the immune and neuroendocrine systems than their luminal counterparts. For this reason, comparative bacteriological analyses were done on rectal biopsies from patients with active colitis, and individuals who had no inflammatory bowel disease. Complex bacterial communities colonized the rectal mucosa in all subjects and great interindividual variabilities in mucosal bacterial populations were observed in both groups. These organisms often occurred in microcolonies, which may have implications for UC, since it would result in higher localized concentrations of bacterial antigens, or toxins, than would be the case if the organisms were diffusely spread across the mucosa.

2004 Inflammatory bowel disease — crossroads of microbes, epithelium and immune systems. Wiley, Chichester (Novartis Foundation Symposium 263) p 57–70

Ulcerative colitis (UC) is one of the two main forms of idiopathic inflammatory bowel disease (IBD), and is generally thought to occur as the result of a genetically determined but abnormal immune response to bacterial species in the normal gut microbiota. A number of investigations have shown that bacteria are important in UC aetiology (Hill 1986, Cummings et al 2003), and several species such as bacteroides, shigellas, streptococci, fusobacteria and desulfovibrios have variously been linked to the disease (Onderdonk 1983, Gibson et al 1993),

[1]This paper was presented at the symposium by E. Furrie, to whom correspondence should be addressed.

however, no specific pathogen has been identified. In some studies (Seksik et al 2003), higher numbers of facultative anaerobes have been observed in faeces from IBD patients, whereas others have reported that antibiotics targeting obligate anaerobes prevented ulceration in animal models (Onderdonk & Bartlett 1979). There is also increased antibody production against anaerobes in UC (Monteiro et al 1971) but no real evidence for a specific transmissible agent (Victor et al 1950). Nevertheless, it is likely that mucosal bacteria are involved in UC, since these organisms exist in close proximity to host tissues and are likely to interact to a greater extent with the host immune system than luminal species.

Bacterial biofilms

Bacteria form biofilms wherever surfaces are available, and they are now thought to be the principal form of microbial growth in natural environments. Biofilms are usually microbial consortia that develop in response to the chemical composition of the substratum, and other biotic and environmental factors. Bacterial biofilms in the large gut give rise to many different microhabitats and metabolic niches. These microenvironments are constantly changing as new resources become available and are consumed. Close spatial relationships between bacterial cells growing on the mucosal surface are important, particularly in relation to metabolic communication between individual species in the microbiota, and between these organisms and the host.

Epithelial populations are difficult to investigate in healthy people and this has restricted study of these microbial communities. Consequently, little is known of their metabolic and health significance, however, they appear to be complex multi-species consortia, and may be highly evolved assemblages, similar to those seen in oral biofilms (Kolenbrander 1989). Intestinal epithelia are covered in a mucus coating 100–200 μm thick (Pullen et al 1994), which probably plays a role in stabilizing bacterial communities associated with the mucosa (Savage 1978). Moreover, the structure and composition of mucosal biofilms will be affected by many host, environmental and microbiological factors, such as innate and adaptive immunity, rates of synthesis and chemical compositions of mucus, turnover rates of intestinal epithelial cells, availability of adhesion sites, lysozyme, gut motility, drugs, disease and bacterial colonization resistance.

Analysis of complex bacterial communities in the gut

Until recently, culturing was the principal method used to identify and quantitate intestinal microorganisms, however, for various reasons, not all bacteria can be cultivated. This can result in underestimation of bacterial population sizes and microbiota diversity, or in some cases, overestimation of the numbers of some important groups of gut bacteria, such as bacteroides and bifidobacteria (Dore et al 1998, Sghir et al 2000). Anaerobic bacteria outnumber aerobes and facultative anaerobes in the gut by a factor of 100–1000 (Simon & Gorbach 1984), and this necessitates the use of anoxic conditions to grow these organisms, together with specialized protocols and apparatus, such as roll tubes, agar shakes, anaerobic jars and chambers (Levett 1991). Because of the complexity of the gut ecosystem, the use of selective agars to detect all of the constituent species is a considerable task, which requires large amounts of consumables and time. Moreover, selective agars are usually not entirely selective, while bacterial species can grow better on some than others. Many bacteria have complex nutritional requirements that are not met by synthetic culture media, or the organisms are obligate syntrophs that will not grow in isolation, and require the metabolic activities or products of other species for their existence (Macfarlane & Gibson 1994). Once isolated, bacteria need to be identified by the use of biochemical tests (Holdeman et al 1977), chemotaxonomic, molecular or immunological analyses (Levett 1991). This often limits the number of samples that can be handled in an investigation, and usually confines identification to the predominant species in the microbiota.

In addition to culturing, a variety of molecular methods of analysis are now available for characterizing bacterial communities in the gut, such as northern hybridizations (Hopkins & Macfarlane 2003), denaturing gradient gel electrophoresis (DGGE, Akkermans et al 2000), real-time PCR (Malinen et al 2003) and FISH (fluorescent *in situ* hybridization), in conjunction with 16S rRNA oligonucleotide probes (Macfarlane et al 2000). However, none of these methods are without limitations, ranging from lack of sensitivity and limited probe availability (northern hybridizations, FISH) to being non-quantitative (DGGE).

Microscopic examination of mucosal biofilms in the large gut

Studies on colonic biopsies from the healthy gut have shown that although mucosal bacteria are able to adhere directly to the bowel wall, the organisms are present in higher numbers in the mucus layer (Hartley et al 1993). These studies indicated that one *E. coli* strain predominated in each tissue sample, and that this strain occurred throughout the length of the gut. Croucher et al (1983) studied colonic tissue from different regions of the bowel in four sudden death victims, none of whom had UC, and noted the existence of distinct bacterial communities in each individual. Due to its tenuous adhesion to the epithelium, the mucus layer and its bacterial populations are often lost during sample preparation, leaving only organisms that are adhering to the epithelial surface. Some workers have reported that the mucus layer in rectal biopsies is more heavily colonized by bacteria in IBD (Schultsz et al 1999). Our studies showed that bacteria occurred throughout the mucus layer, but not in healthy crypts, and that many organisms, such as bacteroides and bifidobacteria, and enterococci in UC patients often grew in microcolonies, especially in the rectum. Live/dead staining of these structures indicated that the majority of these organisms were living, suggesting that they were actively growing on the gut surface. Growth of highly immunogenic bacterial species in microcolonies may be of significance in UC, due to higher localized concentrations of bacterial antigens.

Composition of mucosal biofilms in healthy people and UC patients

Because of the potential role of mucosal biofilms in UC aetiology, we studied these bacterial communities in colitis patients. Biopsies were taken from 20 subjects, nine of which had active UC. Bacteria were characterized using a chemotaxonomic approach, based on their membrane fatty acid methyl ester profiles.

Biofilm composition on the rectal mucosa was highly complex in UC patients and individuals who had no symptoms of IBD. Bacterial cell population densities were in the region of 10^6–10^7 bacteria per cm^2. Biofilm composition varied markedly in different individuals. Microbial communities on the rectal mucosa differed from those in the gut lumen in that facultative anaerobes occurred in proportionately much higher numbers on the mucosa. The predominant facultative species was *Escherichia coli*. This organism has been linked to UC by several workers, although some studies have found that it is uncommon in UC tissues (Walmsey et al 1998).

Bacteroides and bifidobacteria occurred in the highest numbers on the rectal mucosa. Great diversity was seen in the bacteroides, with 16 different species being identified. Other investigators have reported major reductions in lactobacilli and bifidobacteria on the gut mucosa in IBD patients, with increased bacteroides (Pathmakanthan et al 1999). The authors inferred that the reduction in lactobacilli in UC enabled bacteroides, as well as other potentially pathogenic species, to establish on the epithelium. However, no link between these bacteria was apparent in our studies.

Unlike other numerically important organisms on the rectal mucosa, bifidobacteria were present in significantly lower numbers (30-fold) in UC patients. Only *Bifidobacterium angulatum* and *Bif. bifidum* were found in both healthy people and UC patients. Low numbers of bifidobacteria, or the absence of particular bifidobacterial species on the mucosa may be of significance in UC, since some of these bacteria have strong immunomodulatory qualities (Ruseler van Embden et al 1994, Famularo et al 1997), and are thought to contribute towards host defences through interactions with the immune system (Lu & Walker 2001). This concept is supported by the results from a synbiotic RCT carried out in our laboratory, in which a bifidobacterial isolate from a healthy rectal mucosa was fed to UC patients, in combination with fructo-oligosaccharides for one month. Several immunological parameters were investigated. Results showed that in the test patients only, synthesis of human beta defensins 2 and 4, which are key inflammatory markers, was reduced 26- and 81-fold, respectively, while interleukin (IL)1α, tumour necrosis factor (TNF)α and IL6 were reduced 15-, 5- and 22-fold. In the most ill patients, C-reactive protein fell to normal levels. The down-regulation in inflammatory indicators correlated with increased bifidobacterial colonization of the rectal mucosa, and reduced sigmoidoscopy scores.

A greater diversity of anaerobic Gram-positive cocci was observed in UC patients, particularly peptostreptococci. Despite the fact that these bacteria are normal members of the faecal microbiota, other studies have also found that these bacteria only occur on mucosal surfaces in UC (Matsuda et al 2000).

Real-time PCR was used to study the long-term stability of mucosal bacterial populations (bifidobacteria, enterobacteria, peptostreptococci, desulfovibrios) in UC patients over a period of 12 months. The results showed that numbers of these bacteria could vary over several log values, and that these bacterial population changes occurred independently of each other. Other measurements using real-time PCR indicated that increasing mucosal bacterial cell population densities were linked with increased sigmoidoscopy scores in UC patients.

Sulphate-reducing bacteria

Several studies have linked dissimilatory sulphate-reducing bacteria (SRB) to UC, but whether these organisms are involved in disease aetiology, or are just able to adapt to environmental changes in the gut (tissue destruction, diarrhoea) is unclear. However, sulphide is thought to be the most important cytotoxic bacterial metabolite, because it increases mucosal permeability and cellular proliferation, while inhibiting butyrate oxidation in epithelial cells, particularly in UC patients (Roediger et al 1997, Babidge et al 1998). Desulfovibrios are the predominant SRB in the human large intestine (Gibson et al 1988) and apart from their links with UC, they are involved in a number of important physiological processes, such as hydrogen disposal. We therefore developed a real-time PCR assay for studying mucosal desulfovibrios in UC patients.

Desulfovibrios occurred in high numbers in rectal biopsies, in health and disease. Typical counts were in the region of 10^5–10^6 per cm^2. While UC patients usually had higher numbers of SRB than healthy people, the differences were not statistically significant, and no real differences were observed in males and females. The fact that healthy people had high numbers of mucosal SRB indicates that if they are involved in UC, a defect in sulphide detoxification in the mucosa is probably needed for disease to occur. Once inflammatory processes have been initiated, or the mucosal antibacterial response compromised, SRB may also produce antigens that damage the epithelium (Zinkevich & Beech 2000).

In conclusion, complex bacterial populations colonize mucosal surfaces in the human large intestine, and differences in community structure are observed in UC. Sessile microorganisms growing in biofilms often behave differently from their free-living, or planktonic forms, for example, their metabolism changes, while they exhibit greater resistance to antibiotics and other environmental factors that are harmful to planktonic cells (Anwar et al 1990, Van Loosdrecht et al 1990, Mozes & Rouxhet 1992). If bacteria growing in mucosal biofilms are responsible for UC, this may in part explain why antimicrobial substances seem to have been ineffective in treating the disease.

References

Akkermans ADL, Zoetendal EG, Favier CF 2000 Temperature and denaturing gradient gel electrophoresis analysis of 16S rRNA from human faecal samples. Biosci Micro 19:93–98

Anwar H, Dasgupta MK, Costerton JW 1990 Testing the susceptibility of bacteria in biofilms to antibacterial agents. Antimicrob Agents Chemother 34:2043–2046

Babidge W, Millard S, Roediger WEW 1998 Sulfides impair short chain fatty acid beta oxidation at acyl-CoA dehydrogenase level in colonocytes: implications for ulcerative colitis. Mol Cell Biochem 181:117–124

Croucher SC, Houston AP, Bayliss CE, Turner RJ 1983 Bacterial populations associated with different regions of the human colon wall. Appl Environ Microbiol 45:1025–1033

Cummings JH, Macfarlane GT, Macfarlane S 2003 Intestinal bacteria and ulcerative colitis. Curr Issues Intest Microbiol 4:9–20

Dore J, Sghir A, Hannequart-Gramet G, Corthier G, Pochart P 1998 Design and evaluation of a 16S rRNA-targeted oligonucleotide probe for specific detection and quantification of human fecal *Bacteroides* populations. Syst Appl Microbiol 21:65–71

Famularo G, Moretti S, Marcellini S, De Simone C 1997 Stimulation of immunity by probiotics. In: Fuller R (ed) Probiotics: therapeutic and other beneficial effects. Chapman & Hall, London, p 133–161

Gibson GR, Macfarlane GT, Cummings JH 1988 Occurrence of sulphate-reducing bacteria in human faeces and the relationship of dissimilatory sulphate-reduction to methanogenesis in the large gut. J Appl Bacteriol 65:103–111

Gibson GR, Macfarlane S, Macfarlane GT 1993 Metabolic interactions involving sulphate-reducing and methanogenic bacteria in the human large intestine. FEMS Microbiol Ecol 12:117–125

Hartley MG, Hudson MJ, Swarbrick ET, Gent AE, Hellier MD, Grace RH 1993 Adhesive and hydrophobic properties of *Escherichia coli* from the rectal mucosa of patients with ulcerative colitis. Gut 34:63–67

Hill MJ 1986 The possible role of bacteria in inflammatory bowel disease. Curr Con Gastroenterol 3:10–14

Holdeman LV, Cato EP, Moore WEC 1977 Anaerobe laboratory manual 4th edn. Virginia Polytechnic Institute and State University, Blacksburg

Hopkins MJ, Macfarlane GT 2003 Non-digestible oligosaccharides enhance bacterial colonization resistance against *Clostridium difficile* in vitro. Appl Environ Microbiol 69:1920–1927

Kolenbrander PE 1989 Surface recognition among oral bacteria: multigeneric coaggregations and their mediators. Crit Rev Microbiol 17:137–159

Levett PN 1991 Anaerobic microbiology. Oxford University Press, Oxford

Lu L, Walker A 2001 Pathologic and physiologic interactions of bacteria with the gastrointestinal epithelium. Am J Clin Nutr 73:1124S–1130S

Macfarlane GT, Gibson GR 1994 Metabolic activities of the normal colonic flora. In: Gibson SAW (ed) Human health: the contribution of microorganisms. Springer Verlag, London, p 17–52

Macfarlane S, Hopkins MJ, Macfarlane GT 2000 Bacterial growth and metabolism on surfaces in the large intestine. Microb Ecol Health Dis 2:64–72

Malinen E, Kassinen A, Rinttila T, Palva A 2003 Comparison of real-time PCR with SYBR Green 1 or 5′-nuclease assays and dot-blot hybridisation with rDNA-targeted oligonucleotide probes in quantification of selected faecal bacteria. Microbiology 149: 269–277

Matsuda H, Fujiyama Y, Andoh A, Ushijima T, Kajinami T, Bamba T 2000 Characterization of antibody responses against rectal mucosa-associated bacterial flora in patients with ulcerative colitis. J Gastroenterol Hepatol 15:61–68

Monteiro E, Fossey J, Shiner M, Draser BS, Allison AC 1971 Antibacterial antibodies in rectal and colonic mucosa in ulcerative colitis. Lancet 1:249–250

Mozes N, Rouxhet PG 1992 Influence of surfaces on microbial activity. In: Melo LF, Bott TR, Capdeville B (eds) Biofilms-Science and Technology. Kluwer Academic Publishers, Doordrecht, p 125–136

Onderdonk AB 1983 Role of the intestinal microflora in ulcerative colitis. In: Hentges DJ (ed) Human intestinal microflora in health and disease. Academic Press, New York, p 481–483

Onderdonk AB, Bartlett MD 1979 Bacteriological studies of experimental ulcerative colitis. Am J Clin Nutr 32:258–265

Pathmakanthan S, Thornley JP, Hawkey CJ 1999 Mucosally associated bacterial flora of the human colon: quantitative and species specific differences between normal and inflamed colonic biopsies. Microb Ecol Health Dis 11:169–174

Pullen RD, Thomas GAO, Rhodes M, Newcombe RG, Williams GT, Allen A, Rhodes J 1994 Thickness of adherent mucus gel on colonic mucosa in humans and its relevance to colitis. Gut 35:353–359

Roediger WEW, Moore J, Babidge W 1997 Colonic sulfide in pathogenesis and treatment of ulcerative colitis. Dig Dis Sci 42:1571–1579

Ruseler van Embden JGH, Schouten WR, van Lieshout LMC 1994 Pouchitis: result of microbial imbalance? Gut 35:658–664

Savage DC 1978 Factors involved in colonization of the gut epithelial surface. Am J Clin Nutr 31:S131–S135

Schultsz C, van den Berg FM, Ten Kate FW, Tytgat GNJ, Dankert J 1999 The intestinal mucus layer from patients with inflammatory bowel disease harbors high numbers of bacteria compared with controls. Gastroenterology 117:1089–1097

Seksik P, Rigottier-Gois L, Gramet G et al 2003 Alterations of the dominant faecal bacterial groups in patients with Crohn's disease of the colon. Gut 52:237–242

Sghir A, Gramet G, Suau A, Rochet V, Pochart P, Dore J 2000 Quantification of bacterial groups within human fecal flora by oligonucleotide probe hybridization. Appl Environ Microbiol 66:2263–2266

Simon GL, Gorbach SL 1984 Intestinal flora in health and disease. Gastroenterology 86:174–193

Van Loosdrecht MCM, Lyklema J, Norde W, Zehnder AJB 1990 Influence of interfaces on microbial activity. Microbiol Rev 54:75–87

Victor RJ, Kirsner JB, Palmer W 1950 Failure to induce ulcerative colitis experimentally with filtrates of feces and rectal mucosa. Gastroenterology 14:398–400

Walmsey RS, Anthony A, Slim R, Pounder RE, Wakefield AJ 1998 Absence of *Escherichia coli*, *Listeria monocytogenes*, and *Klebsiella pneumoniae* antigens within inflammatory bowel disease tissues. J Clin Pathol 51:657–661

Zinkevich V, Beech IB 2000 Screening of sulfate-reducing bacteria in colonoscopy samples from healthy and colitic gut mucosa. FEMS Microbiol Ecol 34:147–155

DISCUSSION

Jewell: Even though the microcolonies are much more marked in the rectum than the right colon, are the species of bacteria the same?

Furrie: We tend to see more of the facultative organisms. The closer to the outside we go, the greater the number of facultative bacteria. The FISH work is not that easy to quantify. It gives a nice snapshot and a pretty picture. It is tantalizing because we can see the bacteria and the host. The idea of looking at the bacteria and host factors actually interacting is something we have been trying to do for a long time. The only way we can quantify bacterial numbers is with real-time PCR (rtPCR), and we haven't finished this analysis. We now have access to emergency bowel surgery specimens: we get around 200 patients who come in for this surgery each year, and we can take serial biopsies down from the terminal ileum down to the rectum. This is what we are now starting to do. The problem with rtPCR is that it is only as good as your primer pairs. To get very specific primers for specific species is very difficult, particularly for bifidobacteria. We have now started looking at gene expression in the bacteria rather than just 16S rDNA, and we are trying to match gene expression and virulence factor expression with bacterial numbers.

Sartor: You concentrated on the colon. You didn't present any data on the small bowel, yet we know that with proximal diversion, there does seem to be an influence on luminal contents (if not bacteria) in the small bowel with Crohn's disease. Have you had an opportunity to look at the small bowel mucosa?

Furrie: We have just started this now that we have access to the material. We now have ethical committee approval and a new clinician on site who is interested in Crohn's disease, and we are hoping to start looking in Crohn's disease in association with him.

Sartor: You emphasized adherence, but did you have the opportunity to look at invasion with FISH?

Furrie: We have done some invasion studies and the data are not very conclusive. For the enterococci we are just starting to develop a range of primers for looking at virulence factors for attachment and invasion. We are also starting to model attachment and invasion with and without mucus, and various other parameters. The gene sequences are available to analyse these virulence factors for some of the colonizing organisms overly represented in UC.

Meddings: You showed beautiful changes in populations with UC and normals. The age old question is whether these changes are cause or effect. Have you looked at radiation proctitis and other non-inflammatory pathologies to compare them?

Furrie: That is the big question. We are trying to answer this in various ways. We are trying to look at patients who present very early on in disease to see whether we can see a progression or change as they go through the disease. The problem with looking at other inflammatory conditions is once again one of access.

Meddings: Patients with radiation proctitis should be pretty easy to get.

Cummings: It is also a question of how much we can do in the lab at one time.

Furrie: The thing about the UC data I showed is that they are all culture data. It takes a very good anaerobic microbiologist a week to process two samples and then several weeks to identify all the organisms. RtPCR has revolutionized how we can rapidly analyse multiple samples for numbers of specific organisms. To culture 10 normals and 10 UCs, and identify all those organisms takes a lot of time and resources.

Parkos: One of the histological correlations seen in IBD is mucus depletion. You mentioned that mucus is a food source for these bacteria.

Furrie: It's a double-edged sword because you lose mucus but you gain blood. You replace one nutrient source with another. How many of those organisms are there because the environment has changed? We haven't addressed mucus depletion directly. It is not the easiest thing to work with. The biopsies we get are miniscule.

Parkos: You also mentioned that there are these microcolonies. Are they true biofilms?

Furrie: Yes, I'd say so.

Parkos: One way you might think of this is that these biofilms persist with the aberrant organism, thereby perpetuating epithelial inflammation.

Furrie: Absolutely, you are driving something that isn't there in the normal situation. That's why the probiotics should be able to reverse it, because you can start to replace some of the undesirable organisms with those that you would rather have on your epithelium.

Stange: We have shown that HBD2 and HBD3 are induced in UC as opposed to Crohn's, which is characterized by a relative defensin deficiency (Wehkamp et al 2003). It is good to see that you also observe a correlation with the sigmoidoscopic score in UC. This probably shows that in that disease, defensin induction is a function of inflammation. With probiotics there are two possibilities, and we have seen both. One is that using the VSL3 from pouchitis we see exactly the same as you have shown with your bifidobacteria: inflammation is reduced, and because of this, expression of the defensins HBD2 and 3 is decreased—most likely through diminished NF-κB and AP1. This is a secondary phenomenon. We have also seen the opposite: the probiotic *E. coli* Nissle both in cell culture as well as in a mouse model induces β-defensins. So, if you have a situation where there is active inflammation, this is shut down and defensin formation goes down in parallel. However, if you have a UC patient in remission, there are studies showing that this can be maintained (Kruis et al 1997, Rembacken et al 1999) where the opposite happens: defensins are induced and this is what prevents relapse.

Furrie: With *E. coli* I am not sure where in the GI tract the defensins are being induced. Would this be more of a small intestine induction than a colonic induction?

Stange: In the mouse it is in the colon; in the cell culture it's a caco-2 cell.

Furrie: We have done similar work with caco-2 and HT29 cells with our probiotic. This doesn't induce defensins. If we have a system where we induce defensin expression with IL1α, we can show a dampening of this using the probiotic on HT29s and caco-2 cells. There is a mechanism for the damping down if you believe cell lines act in a similar manner to primary tissue. For our UC studies we measure expression in rectal biopsies. We have no idea what is happening further up the bowel and in the small intestine. There is evidence that prebiotics might be affecting things quite markedly in the small intestine.

Rhodes: It is fascinating that you are finding changes in the rectum. People have had difficulty explaining the rectal predominance of UC. Is your biofilm underneath the mucus layer? Do you wash the mucus off?

Furrie: There is no treatment of these patients at all. When we analyse samples with confocal microscopy we find that there are bacteria growing through the mucus as well as sitting directly on the epithelial cells.

Rhodes: Can you identify the ones on the epithelium?

Furrie: Yes, using specific 16S RNA probes we can identify enterococci at that site.

Rhodes: Are they able to get through the mucus? I'd always imagined that they would need flagellae to do this.

Furrie: It gets back to the issue of mucus depletion and replacement with the bloody, inflamed bowel.

Rhodes: So these may be on the mucosa in situations where the mucus layer is defective or absent.

Furrie: Yes. If you model colon bacterial population in a chemostat and introduce mucus baits into this system, the bifidobacteria will effectively eat their way through the mucus and arrive at the bottom of the cup where you could envisage the epithelial cells would be. They will digest their way through the mucus and settle on the epithelial cells. We have nice confocal images of bifidobacteria and bacteroides where the bifidobacteria have penetrated the mucus and have moved through the layer but the bacteroides remain on the top of the mucus bait.

Roediger: You mentioned redox control, and a 'chicken and egg situation'. Can you measure redox control in the early phases of disease? Can you measure it at all?

Furrie: I don't know how we could measure this in these biopsies. The only indication of this is the replacement of obligate anaerobes with facultative bacteria.

Fedorak: I have a methodological question. Are the patients in whom you are doing colonoscopy and obtaining biopsies prepped with a colonic lavage?

Furrie: No.

Fedorak: In that regard your results represent data from unprepped colonoscopy and unprepped biopsies. Do you think the colonic lavage preparation would change the bacterial composition within the lumen or adherent to the mucosa?

Furrie: I would expect so, although we haven't measured this. Studies on prepped samples show different numbers than we see in our studies (Swidsinski et al 2002). We specifically don't prep. It was a decision made because we really wanted to maximize what we could see on the epithelium. A lot of work has been done on luminal contents, and luminal bacterial numbers and species diversity appears to be entirely different in prepped and unprepped studies.

Fedorak: In obtaining mucosal biopsies only you may be missing an important contribution of luminal bacteria. Is there any evidence for a role of luminal bacteria?

Furrie: We don't know that. The reason we are focusing on the mucosa is for diseases such as inflammatory bowel disease. You'd imagine that these organisms are interacting directly with the host. What would happen with luminal organisms producing toxic products? We haven't addressed this. We have done some other feeding studies in which we have looked at luminal populations in healthy ageing, frail elderly and in antibiotic-treated hospitalized elderly. We do see quite dramatic differences in luminal populations with factors such as transit time and antibiotic treatment. We haven't asked this question in UC.

Peña: What is your rationale for using a combination of pre- and probiotics?

Furrie: This was to give the probiotic the best chance of colonizing. We screened about 50 different bifidobacterial isolates from healthy mucosa, for a whole variety of factors including bile acid and oxygen tolerance. We selected this organism because it was the most robust and also because it grew incredibly well with a particular probiotic which is an inulin/fructo-oligosaccharide.

Peña: When you stopped treatment did it persist in the gut?

Furrie: The people who have gone into remission have generally stayed in remission. The study just finished two months ago, but we have been doing this for about two years. We haven't had any dramatic relapses.

Peña: Is the bifidobacterium still clearly detectable?

Furrie: We are analysing this. All I can say now is that we see an increase in bifidobacteria on the mucosa. We are trying to develop more species-specific primer pairs, but this is proving to be incredibly difficult. Bifidobacteria are a complex group of organisms but there are only small differences in different strains of the same species.

Bjarnason: We are all accepting that there is some symbiosis between the bacteria in the gut and the mucosa. If I take the example of transgenic mice, they are all perfectly happy in germ-free conditions. If we take a germ free normal animal it has beautiful villi and no inflammation. The price of eating and not being in a germ free environment is that it gets subclinical inflammation in the gut so it has low-grade IBD. If this is the price we pay for a gut flora, what is the benefit we get from them? I suspect it is precious little.

Furrie: The epithelial cells are delighted because they get butyrate, a main energy source.

Bjarnason: You can have many generations of germ-free animals that are perfectly happy without this energy source.

Furrie: Germ-free animals are not germ-free humans. The human gut is different.

Bjarnason: I suspect if I could sterilize my gut with malt whisky, then I would be happy! Low grade IBD is a terrible price to pay for having a gut flora, as I can't see any benefit from it.

Rhodes: We are getting into the realm of philosophy here. I thought the argument was that if you are close to starvation, you do get considerable calories from breaking down the fibre in the colon.

Sartor: Is that true in humans?

Gibson: Yes, we get 10% of our calories from fermentation.

Bjarnason: If it is a question of energy source then it should be obvious that if I sterilized my gut I would lose some weight! It seems to me I get little benefit from these bacteria.

Schölmerich: We know that bacteria regulate a lot of genes in our epithelia and our immune system which we might need for other functions.

Bjarnason: Why is that not part of the IBD that you have as a normal individual in response to bacteria? This could equally be interpreted as a reaction to the bacteria being there.

Schölmerich: They don't regulate just inflammatory or anti-inflammatory genes, but also colipase for example.

Bjarnason: Without these bacteria present we wouldn't need the white cells there.

Uhlig: We did a similar study to that just presented by Elizabeth Furrie in a mouse model of colitis. In the T cell transfer model of colitis we controlled the onset of disease and compared inflammatory and non-inflammatory conditions. Under the non-inflammatory conditions we saw that the number of directly adherent bacteria was very low. A mucus film separates the bacteria from the epithelium. There was a depletion of mucus when the inflammation started, and different bacterial species became adherent. We interpreted the increased bacterial adherence as a secondary effect due to the inflammation, and the inflammation induced lack of mucus as the primary event.

Furrie: I think this may be true for humans: where most of the changes in bacterial species in UC are secondary to the inflammation. But I think the question about what is driving IBD is still open. We would say that colonization by these organisms that are not normally present drives IBD, although this is probably not the initiating signal.

Uhlig: What therapy were these patients on?

Furrie: They were kept on the therapy that they were on when they entered the trial. They are a mixed bag, but we wanted to maximize our population and didn't want to be selective about who we can treat. We'd like to put the capsules in the post and send them to people all over the UK.

Kelleher: Are there any particular effects of these bifidobacteria on pattern recognition receptors.

Furrie: We have developed a large number of primer pairs for many immune receptors, and also defensins and various cytokines. On initial analysis of all Toll-like receptors we don't see a huge change. We have just started looking at histology rather than trying to do the blanket up-and-down regulation of genes. We are using rtPCR as a screening step, and we are now starting to look at protein expression through immunohistochemistry in these patients. Now that decent monoclonal antibodies are available we can look at expression patterns and cell types.

Gibson: I believe *Clostridium butyricum* is being used as a probiotic in the USA. Is it present?

Furrie: Yes, although not in huge numbers.

Gibson: I'm surprised I haven't heard a whisper about desulfovibrios. Are they present?

Furrie: Yes, we can see them with FISH probes. They are on the mucus generally. We keep trying to find them on the mucosa, but with FISH we are limited by cell number, because there is a limit to what you can see with the fluorescence. There needs to be a minimum number of cells that are reasonably active with quite a lot of 16S rRNA. With RT-PCR we can detect them (Fite et al 2004).

Gibson: Are they in UC but not controls?

Furrie: No, they are in both.

Sartor: I was fascinated by the sterile nature of the crypts in the normal patients. Given the up-regulation of aerobic bacteria in UC is the crypt still sterile?

Furrie: The problem with the UC samples is that there is very little normal colonic morphology left in bad cases, and the biopsies are very small so we can't tell.

References

Fite A, Macfarlane GT, Cummings GH et al 2004 Identification and quantitation of mucosal and faecal desulphovibrios using real time PCR. Gut 53:523–529

Kruis W, Schutz E, Frie P et al 1997 Double-blind comparison of an oral Escherichia coli preparation and mesalazine in maintaining remission of ulcerative colitis. Aliment Pharmacol Ther 11:853–858

Rembacken BJ, Snelling AM, Hawkey PM et al 1999 Non-pathogenic Escherichia coli versus mesalazine for the treatment of ulcerative colitis: a randomised trial. Lancet 354:635–639

Swidsinski A, Ladhoff A, Pernthaler A et al 2002 Mucosal flora in inflammatory bowel disease. Gastroenterology 122:44–54

Wehkamp J, Harder J, Weichenthal M et al 2003 Inducible and constitutive beta-defensins are differentially expressed in Crohn's disease and ulcerative colitis. Inflamm Bowel Dis 9:215–223

Defensins and mucosal protection

Y. R. Mahida and R. N. Cunliffe

Division of Gastroenterology & Institute of Infection, Immunity and Inflammation, University of Nottingham and University Hospital, Queen's Medical Centre, Nottingham NG7 2UH, UK

Abstract. The gastrointestinal tract is constantly exposed to microorganisms, mainly a large and complex population of bacteria resident in the colon and distal small intestine. Although the normal host relationship with the resident luminal bacteria is often mutually beneficial, the host also requires protection against these microorganisms. Epithelial cells play a critical role in mediating these protective responses and there is increasing appreciation of the likely importance of antimicrobial peptides of the defensin family that they express. The enteric α-defensins (human defensins [HD]5 and 6) are expressed by Paneth cells, which are normally confined to the small intestine, but are also seen in the colon in patients with inflammatory bowel disease. Studies have shown that HD5 is stored in Paneth cell granules in precursor form and requires processing to the mature peptide. Human β-defensin (HBD)1 is constitutively expressed in intestinal epithelial cells, whereas the expression of HBD2 is induced in inflammatory bowel disease. HBDs have also been shown to be chemotactic for immature dendritic cells and memory T cells. Thus, they may not only mediate innate immunity, but also regulate adaptive immune responses in inflammatory bowel disease.

2004 Inflammatory bowel disease — crossroads of microbes, epithelium and immune systems. Wiley, Chichester (Novartis Foundation Symposium 263) p 71–84

In the normal intestine there is a highly complex community of bacteria in the colon and distal small intestine, which according to one estimate, outnumbers the total number of cells in the host by a ratio of 10:1 (Savage 1977). The host mucosal-resident microbial interactions are ill understood but some of the molecular aspects are beginning to be characterized (Hooper et al 2002). Additional exposure of the intestine to microorganisms, including pathogens, occurs via the oral route. Significant current research effort is directed towards the understanding of mechanisms by which the normal intestinal mucosa is able to establish a symbiotic relationship with the commensal flora, while retaining the capacity to respond to pathogenic microorganisms via an acute inflammatory response.

There has been increasing appreciation of the importance of luminal bacteria in the pathogenesis of inflammatory bowel disease. Initial studies in patients provided important clues regarding the contribution of the resident luminal bacteria in the disease process (Harper et al 1985). Studies over the last decade in rodent models have illustrated the importance of interactions between host mucosal cells and resident luminal microorganisms in the development and perpetuation of chronic intestinal inflammation in which there are histological similarities to those seen in patients with inflammatory bowel disease (Strober et al 2002). The surface monolayer of epithelial cells is constantly in close proximity to the luminal micoorganisms and their products and the functions of the enterocytes are critical to host defence. These functions include the expression of antimicrobial peptides and proteins.

The intestinal epithelium comprises distinct subpopulations of cells derived from stem cells in the crypt. In the small intestine four main epithelial cell types are recognized, absorptive enterocytes, goblet cells, enteroendocrine cells and Paneth cells. The latter are usually absent in the normal colon but are seen in patients with inflammatory bowel disease. In contrast to the other epithelial cell types of the normal small intestine, Paneth cells are present in the crypt base and are long-lived. These granulated cells have generated considerable interest because they express a number of antimicrobial peptides and proteins (Ouellette & Bevins 2001), which include lysozyme, secretory phospholipase A2 and enteric α-defensins. Recent studies have shown that intestinal epithelial cells may also express members of the β-defensin family. Defensin peptides contain 3 disulphide bonds formed from 6 cysteine residues. They are divided into α- and β-defensins based on the location of the cysteine residues and pairing of disulphide bonds (Cunliffe & Mahida 2004).

Enteric α-defensins

Human neutrophil defensins were described in the 1980s and enteric α-defensins were first identified in the murine small intestine (designated cryptdins). In humans, two forms of enteric α-defensins have been identified, human defensin (HD)5 and 6 (Cunliffe & Mahida 2004). Studies to date suggest that in the human and murine gastrointestinal tract, expression of enteric α-defensins is restricted to Paneth and related cells. Thus, HD5 is expressed in normal small intestinal Paneth cells and a small unique population of villous epithelial cells with features of both Paneth and goblet cells, designated intermediate cells (Cunliffe et al 2001). In addition to HD5, these latter cells also expressed intestinal trefoil factor (normally a goblet cell-specific protein) but not lysozyme, which is expressed by Paneth cells at the base of small intestinal crypts (Cunliffe et al 2001). HD5 is not expressed in the normal colon (which lacks Paneth cells) but is present in metaplastic Paneth cells in the colon affected by inflammatory bowel disease (Cunliffe et al 2001, Wehkamp et al 2002, Fahlgren et al 2003).

In the normal human and murine small intestinal Paneth cells, the enteric α-defensins are synthesized in precursor form and are subsequently processed to mature peptides. In mice, the enteric α-defensins are processed to active peptides by the matrix metalloproteinase enzyme matrilysin (MMP7; which is also expressed in Paneth cells; Wilson et al 1999), before secretion into the lumen. By contrast, HD5 is stored in precursor form in human Paneth cells and is processed during and/or after secretion (Cunliffe et al 2001). In our studies, antiserum was raised against a chemically synthesized putative mature form of HD5 and was subsequently used for immunohistochemical studies and for purification of HD5 from acid extracts of normal terminal ileal crypts. The major component of HD5 isolated from pooled ileal crypts had a mass and N-terminal sequence similar to that predicted for the precursor form of HD5. When isolated crypts were stimulated with lipopolysaccharide or carbamylcholine chloride, one truncated form of HD5 was identified in the secretions, whose N-terminal sequence suggested partial processing of the precursor form, prior to cleavage to the mature form of the peptide (Cunliffe et al 2001).

In studies by others, trypsin has been implicated in the processing of pro-HD5 to the mature form (Ghosh et al 2002) but based on the findings of our studies (above) we speculate that other processing enzymes may also be involved. Overall, the studies in human and mice have shown that there is a high constitutive level of expression of enteric α-defensins in Paneth cells and the biological activities of these peptides are regulated by processing enzymes and mediators capable of inducing degranulation. The expression of the enteric α-defensins does not require the presence of luminal microorganisms (Putsep et al 2000), although in a recent report, luminal microbial flora induced the expression of antimicrobial angiogenin 4 in germ-free mice (Hooper et al 2003). Oral inoculation of *Salmonella enterica* serovar Typhimurium, a human pathogen in food-borne enteritis, has recently been reported to inhibit the expression of Paneth cell cryptdins (Salzman et al 2003a), illustrating a potential virulence mechanism by such pathogens.

Our studies in mice infected with the nematode *Trichinella spiralis* have revealed another mechanism by which expression of enteric α-defensins may be up-regulated. Following infection with the nematode, there is an increase in the number of Paneth cells and also intermediate cells, which have characteristics of both Paneth and goblet cells (Kamal et al 2001). The Paneth and intermediate cells expressed cryptdin enteric α-defensins but in contrast to human intermediate cells (Cunliffe et al 2001), those in *T. spiralis*-infected mice did not express the goblet cell-specific intestinal trefoil factor (Kamal et al 2001). Following *T. spiralis* infection, there was active degranulation of Paneth cells into the lumen during infection with a return to baseline of Paneth cell numbers and morphology following worm expulsion. The increase in Paneth and intermediate cell numbers appears to be mediated by a unique population of mucosal T cells, implying regulation of innate immunity by the adaptive immune system.

In vitro, HD5 and murine α-defensins express a wide spectrum of antimicrobial activity against both Gram-negative and Gram-positive bacteria. Antimicrobial activities of these peptides *in vivo* have also been demonstrated in normal mice (Wilson et al 1999) and transgenic mice expressing HD5 in Paneth cells (Salzman et al 2003b). The normal restriction of defensin-expressing Paneth cells to the small intestine has led to the suggestion that these cells maintain relative sterility in this region of the gastrointestinal tract but direct evidence for this proposition is awaited. The presence of Paneth cells in the colon in inflammatory bowel disease is likely to reflect enhancement of antimicrobial host defence. Paneth cells arise following differentiation of stem cells in the crypt but factors regulating stem cell differentiation remain to be characterized. Our studies in *T. spiralis*-infected mice suggest an important contribution by T cells during chronic inflammation (see above).

β-defensins in the intestine

An increasing number of β-defensin peptides have been described over recent years and are expressed not only in the gastrointestinal tract, but also the lung, eye and skin. These peptides are synthesized as preproproteins and processed intracellularly to mature forms by mechanisms that are not fully understood. Human β-defensin (HBD)1 is widely and constitutively expressed in epithelial cells throughout the gastrointestinal tract. By contrast there is low level of expression of HBD2 in the normal intestine but its expression is up-regulated in active inflammatory bowel disease (O'Neil et al 1999, Fahlgren et al 2003). This up-regulation in expression of HBD2 is likely to be mediated via nuclear factor (NF)-κB (Takahashi et al 2001, Tsutsumi-Ishii & Nagaoka 2002). In contrast to the enteric α-defensins (HD5 and 6), HBD1 and HBD2 are expressed predominantly by columnar epithelial cells. HBD1 and HBD2 are active against Gram-negative bacteria but have limited activity against Gram-positive bacteria such as *Staphylococcus aureus*. In early infections by *Shigella* spp. the expression of HBD1 and LL-37 (an antimicrobial peptide expressed by epithelial cells) has been reported to be reduced in biopsies of patients with bacillary dysenteries and in *Shigella*-infected cell cultures *in vitro* (Islam et al 2001). Thus, down-regulation of epithelial antimicrobial peptide expression may facilitate bacterial invasion.

An interesting biological activity of HBD1 and HBD2 is their ability to induce chemotaxis of immature dendritic cells and memory T cells, following the direct binding and activation of the chemokine receptor CCR6 (Yang et al 1999). The only other known ligand for CCR6 is macrophage inflammatory protein (MIP)3α, which is expressed by intestinal epithelial cells (Izadpanah et al 2001,

Kwon et al 2002). Thus, up-regulation of HBD2 expression in active inflammatory bowel disease would be expected to enhance adaptive immunological responses via chemotaxis of immature dendritic cells and memory T cells. More recent studies using HBD3 (which has similar chemotactic activities to HBD1 and HBD2) have shown that the disulphide bonds are not required for antimicrobial activity but are essential for chemotactic properties (Wu et al 2003).

Other antimicrobial peptides and proteins

Human neutrophil defensins (HNP) are expressed by polymorphonuclear cells infiltrating the mucosa in active inflammatory bowel disease and immunohistochemical studies suggest secretion of these peptides in the tissue (Cunliffe et al 2002). Normal intestinal epithelial cells do not express HNP but studies by immunohistochemistry and immunoblot analysis after acid urea–polyacrylamide gel electrophoresis have demonstrated the expression of these peptides in epithelial cells of mucosal samples affected by active inflammatory bowel disease (Cunliffe et al 2002). The neutrophil defensins were expressed by scattered columnar epithelial cells, often predominantly in the apical region of the cells. Many HNP-expressing epithelial cells in the inflamed mucosal samples also expressed the antimicrobial protein lysozyme.

We have also isolated histone H1 and its fragments from acid extracts of human terminal ileal mucosal samples (Rose et al 1998). Histone H1 was demonstrated in the cytoplasm of enterocytes at the villus tip of sections of the normal small intestine. Cells in the same position as histone H1-immunoreactive enterocytes have been shown by *in situ* end labelling to be highly susceptible to programmed cell death (Hall et al 1994). It is therefore of interest that recent studies have shown the release of H1 forms of histone into the cytoplasm after irradiation, with one of the H1 forms (H1.2) playing an important role in the transmission of apoptotic signals from the nucleus to the mitochondria following DNA double-strand breaks (Konishi et al 2003). Pathogenic microorganisms may induce apoptosis in intestinal epithelial cells (Mahida et al 1996, Zychlinsky & Sansonetti 1997) and the release of antimicrobial histone H1 may represent a distinct form of host defence to avoid invasion by the pathogen or other luminal bacteria.

Conclusions

There is increasing appreciation of the likely importance of epithelial cell-expressed antimicrobial peptides of the defensin family in host protection against luminal microorganisms and in the pathogenesis of inflammatory bowel disease. Distinct subpopulations of epithelial cells may express α- or β-defensins and while the former are constitutively expressed in Paneth cells, production of some members of the latter family is induced during inflammation. Enteric α-defensin expression is induced in the colon in inflammatory bowel disease via metaplastic Paneth cells. Nematode infection may also induce an increase in the number of Paneth and α-defensin-expressing intermediate cells in the small intestine. The ability of human β-defensins to induce chemotaxis of immature dendritic cells and memory T cells illustrates the capacity of epithelial cell-mediated innate immunity to regulate adaptive immune responses in the intestine in inflammatory bowel disease.

References

Cunliffe RN, Kamal M, Rose FR, James PD, Mahida YR 2002 Expression of antimicrobial neutrophil defensins in epithelial cells of active inflammatory bowel disease mucosa. J Clin Pathol 55:298–304

Cunliffe RN, Mahida YR 2004 Expression and regulation of antimicrobial peptides in the gastrointestinal tract. J Leukoc Biol 75:49–58

Cunliffe RN, Rose FR, Keyte J et al 2001 Human defensin 5 is stored in precursor form in normal Paneth cells and is expressed by some villous epithelial cells and by metaplastic Paneth cells in the colon in inflammatory bowel disease. Gut 48:176–185

Fahlgren A, Hammarstrom S, Danielsson A, Hammarstrom ML 2003 Increased expression of antimicrobial peptides and lysozyme in colonic epithelial cells of patients with ulcerative colitis. Clin Exp Immunol 131:90–101

Ghosh D, Porter E, Shen B et al 2002 Paneth cell trypsin is the processing enzyme for human defensin-5. Nat Immunol 3:583–590

Hall PA, Coates PJ, Ansari B, Hopwood D 1994 Regulation of cell number in the mammalian gastrointestinal tract: the importance of apoptosis. J Cell Sci 107:3569–3577

Harper PH, Lee EC, Kettlewell MG, Bennett MJ 1985 Role of the faecal stream in the maintenance of Crohn's colitis. Gut 26:279–284

Hooper LV, Midtvedt T, Gordon JI 2002 How host-microbial interactions shape the nutrient environment of the mammalian intestine. Annu Rev Nutr 22:283–307

Hooper LV, Stappenbeck TS, Hong CV, Gordon JI 2003 Angiogenins: a new class of microbicidal proteins involved in innate immunity. Nat Immunol 4:269–273

Islam D, Bandholtz L, Nilsson J et al 2001 Downregulation of bactericidal peptides in enteric infections: a novel immune escape mechanism with bacterial DNA as a potential regulator. Nat Med 7:180–185

Izadpanah A, Dwinell MB, Eckmann L, Varki NM, Kagnoff MF 2001 Regulated MIP-3alpha/CCL20 production by human intestinal epithelium: mechanism for modulating mucosal immunity. Am J Physiol Gastrointest Liver Physiol 280:G710–G719

Kamal M, Wakelin D, Ouellete AJ, Smith A, Podolsky DK, Mahida YR 2001 Mucosal T cells regulate Paneth and intermediate cell numbers in the small intestine of T. spiralis-infected mice. Clin Exp Immunol 126:117–125

Konishi A, Shimizu S, Hirota J et al 2003 Involvement of histone H1.2 in apoptosis induced by DNA double-strand breaks. Cell 114:673–688

Kwon JH, Keates S, Bassani L, Mayer LF, Keates AC 2002 Colonic epithelial cells are a major site of macrophage inflammatory protein 3alpha (MIP-3alpha) production in normal colon and inflammatory bowel disease. Gut 51:818–826

Mahida YR, Makh S, Hyde S, Gray T, Borriello SP 1996 Effect of Clostridium difficile toxin A on human intestinal epithelial cells: induction of interleukin 8 production and apoptosis after cell detachment. Gut 38:337–347

O'Neil DA, Porter EM, Elewaut D et al 1999 Expression and regulation of the human beta-defensins hBD-1 and hBD-2 in intestinal epithelium. J Immunol 163:6718–6724
Ouellette AJ, Bevins CL 2001 Paneth cell defensins and innate immunity of the small bowel. Inflamm Bowel Dis 7:43–50
Putsep K, Axelsson LG, Boman A et al 2000 Germ-free and colonized mice generate the same products from enteric prodefensins. J Biol Chem 275:40478–40482
Rose FR, Bailey K, Keyte JW, Chan WC, Greenwood D, Mahida YR 1998 Potential role of epithelial cell-derived histone H1 proteins in innate antimicrobial defense in the human gastrointestinal tract. Infect Immun 66:3255–3263
Salzman NH, Chou MM, de Jong H, Liu L, Porter EM, Patterson Y 2003a Enteric salmonella infection inhibits Paneth cell antimicrobial peptide expression. Infect Immun 71:1109–1115
Salzman NH, Ghosh D, Huttner KM, Patterson Y, Bevins CL 2003b Protection against enteric salmonellosis in transgenic mice expressing a human intestinal defensin. Nature 422:522–526
Savage DC 1977 Microbial ecology of the gastrointestinal tract. Annu Rev Microbiol 31:107–133
Strober W, Fuss IJ, Blumberg RS 2002 The immunology of mucosal models of inflammation. Annu Rev Immunol 20:495–549
Takahashi A, Wada A, Ogushi K et al 2001 Production of beta-defensin-2 by human colonic epithelial cells induced by Salmonella enteritidis flagella filament structural protein. FEBS Lett 508:484–488
Tsutsumi-Ishii Y, Nagaoka I 2002 NF-kappa B-mediated transcriptional regulation of human beta-defensin-2 gene following lipopolysaccharide stimulation. J Leukoc Biol 71:154–162
Wehkamp J, Schwind B, Herrlinger KR et al 2002 Innate immunity and colonic inflammation: enhanced expression of epithelial alpha-defensins. Dig Dis Sci 47:1349–1355
Wilson CL, Ouellette AJ, Satchell DP et al 1999 Regulation of intestinal alpha-defensin activation by the metalloproteinase matrilysin in innate host defense. Science 286:113–117
Wu Z, Hoover DM, Yang D et al 2003 Engineering disulfide bridges to dissect antimicrobial and chemotactic activities of human beta-defensin 3. Proc Natl Acad Sci USA 100:8880–8885
Yang D, Chertov O, Bykovskaia SN et al 1999 Beta-defensins: linking innate and adaptive immunity through dendritic and T cell CCR6. Science 286:525–528
Zychlinsky A, Sansonetti P 1997 Perspectives series: host/pathogen interactions. Apoptosis in bacterial pathogenesis. J Clin Invest 100:493–495

DISCUSSION

Jewell: I have a slightly naïve question. If Paneth cells really are a key factor in protecting us from bacterial damage, and perhaps controlling the number of organisms in the gut, it is surprising that the least number of organisms is found in the proximal intestine where there are no Paneth cells. One might have expected Paneth cells to have been equally distributed up and down the gut. Why are they just found in the terminal ileum? Is it just to protect against the colon organisms spilling back?

Mahida: That is a good question. The proposition that Paneth cells control the normal bowel flora and restrict it to the colon is very attractive. However, definitive evidence for this function of Paneth cells is awaited.

Schölmerich: In the upper gastrointestinal (GI) tract we have bile at high concentrations, in the millimolar range. This interacts strongly with the bacteria and increases gut motility, so they can't really stay there. If you do a biliary drainage, and look for bacterial colonization of the upper GI tract, bacteria are then present in higher numbers.

Kelleher: Why do you think human HD5 has to be processed externally, from an evolutionary point of view, as opposed to the murine homologue? Do you think this is related to differences in colonization in the GI tract?

Mahida: We postulate that the precursor form may have different functions. Bevins and colleagues have found that the precursor form of HD5 has antimicrobial activity against *Listeria monocytogenes*, but not against *Salmonella typhimurium*, against which mature HD5 is active (Ghosh et al 2002). In normal mice, enteric α-defensins are processed to the mature form by MMP7. In transgenic mice expressing the *HD5* gene in Paneth cells, the mature form of the peptide has been demonstrated in the small intestinal lumen (Salzman et al 2003). This begs the question, what is processing the precursor form of HD5 in murine Paneth cells? The answer to this question remains to be addressed.

Kelleher: So the mature form is emerging from the cells, or the mature form is detected in the small bowel. Could it be a bacterial enzyme that is actually processing it?

Mahida: I think this is possible, because there are a number of potential cleavage sites. *In vitro* trypsin can process the precursor to the mature form of HD5 and one can postulate that trypsin present in the lumen could mediate this effect.

Peña: I thought that there were a lot of Paneth cells in the duodenum. For example, lysozyme is very important. I don't know where HD5 is present in humans. Around the 1960s Brian Creamer described a form of total villous atrophy that was unresponsive. He made an issue of the fact that there were no Paneth cells there (Creamer & Pink 1967).

Mahida: We have found HD5-expressing Paneth cells in sections of normal duodenum.

Stange: These are elegant studies. What needs to be followed up in detail is the biochemical array of endogenous antibiotics in the intestine. We don't know the quantitative role of histone, β-defensins and so on. As mentioned before, a key point for explaining Crohn's disease could be that in contrast to UC, there is an induction problem for three β-defensins, HBD2, HBD3 and a new one called psoriasin in the colon. The colon in Crohn's disease seems to defend itself by metaplastic Paneth cells which produce HD5 and 6 to compensate for the deficiency of β-defensins. It is a different story in ileal Crohn's where α-defensins are deficient if there is ileal involvement. With Crohn's disease of the colon only, ileal α-defensins are completely normal. If you have a *NOD2* mutation as well as ileal involvement, then α-defensins are almost completely absent. This suggests that ileal Crohn's is a result of HD5 and HD6 deficiency there. We have also transfected Caco-2 cells with the mutated *NOD2*, as opposed to the control

NOD2. There was clearly deficient induction even of *β*-defensins in that system. NOD2 seems to be involved in the induction of defensins and that is why it is expressed in Paneth cells. Defensin induction in UC is a probably secondary phenomenon to inflammation, but in Crohn's this seems to be the primary defect (Fellermann et al 2003).

Gibson: What actually happens to the defensins once they are secreted? Do they go into the lumen and keep acting? Do they get degraded very rapidly? Where do they work?

Mahida: The mature form of HD5 has been shown to be present in endoscopic aspirates of small intestine lumen. Studies have also been done in ileal neobladder urine—this is where the ileum is constructed into a bladder—and a variety of different forms (including precursor and truncated forms) of HD5 have been identified. One of the truncated forms is identical to the partially processed form of HD5 that we obtained following stimulation of Paneth cell secretion in isolated crypts. The question of whether these different forms have distinct functions is an interesting one. Work on murine α-defensins (cryptdins), has shown that the N-terminus is important in shaping antimicrobial activity. For example cryptdin 4 is highly active against *E. coli* and is expressed at maximal levels in the terminal ileum. By contrast, cryptdin 2 is active against *G. lamblia* but has limited activity against *E. coli* (Ouellette & Selsted 1996).

Gibson: So the terminal ileal Paneth cells are producing antibacterial substances for the colon?

Mahida: This is possible, but the relationship between the expression of antimicrobial peptides and the luminal microbial flora is currently under active investigation. Germ-free mice are able to process enteric α-defensins and in humans, Paneth cells are seen at week 12 of gestation, when HD5 and HD6 are also present. Thus, luminal microorganisms do not seem to be required for expression of human and murine enteric α-defensins. However, in recent studies, the antimicrobial molecule angiogenin 4 has been shown to be induced by luminal microbial flora (Hooper et al 2003).

Parkos: Aren't defensins also regulated by the ionic milieu?

Mahida: This point is particularly relevant for cystic fibrosis (CF). The change in salt concentration in the layer above the epithelium in CF leads to loss of *β*-defensin activity. However, there has been an on-going debate regarding the importance of the depth of the mucus layer and its salt concentration in the pathogenesis of CF. In antimicrobial assays for defensins, salt concentration is an important issue.

Uhlig: If mammals have such a restricted set of defensins, why do defensins still work? If we used a single antibiotic for 10 years we'd be selecting for resistance. It is surprising that resistance hasn't developed to this restricted set of broad spectrum defensins.

Mahida: I am not sure we know the full spectrum of activity *in vivo*. The transgenic HD5 mouse model shows for the first time that HD5 may be important for mediating antimicrobial activity *in vivo*. Almost all the work to date with the defensins has been *in vitro*. This work suggests there may be some antimicrobial peptides that are not as active against the resident flora as other bacteria. It has been postulated that when the mouse intestine acquires its resident flora, angiogenin 4 influences the community of the resident flora that develops subsequently.

Stange: This is work in progress. In the genome there are at least four unknown genes that look like β-defensins, and we only know six up to now.

Kelleher: At what level do HBD1 and 2 impact on the process of chemotaxis? Are they mediating increased levels of CCR6 or are they up-regulating receptor levels? What is the process that is involved in this?

Mahida: I am not aware of studies showing alterations in the expression of CCR6 in response to HBD1 and HBD2. Chemotaxis by these peptides has been reported to be mediated via interaction with CCR6. Other antimicrobial peptides are also capable of inducing chemotaxis of T cells and monocytes. They include human neutrophil α-defensins and LL-37, which is expressed by colonic epithelial cells.

Kelleher: Is there homology with other CCR6 ligands?

Mahida: I am not aware of this.

Furrie: You can actually buy recombinant β-defensins 1, 2 and 3. We have done some work with them looking at their effects on commensal organisms isolated from the mucosa. The bottom line is that the bifidobacteria we have measured aren't affected by defensins but bacteroides species are highly susceptible to killing. They may well affect which organisms can live at the epithelium and which can live above it.

Mahida: The availability of recombinant peptides will make *in vitro* analysis much easier. Hopefully, *in vivo* systems will be developed that will enable us to address their relative importance in protection and disease pathogenesis.

Rhodes: I was thinking about your intriguing parasite model and the role of the T cells in inducing dramatic Paneth cell hyperplasia, trying to link this with the extraordinary focusing of Paneth cells at the bottom of the crypts. Are the relevant T cells, expressing TNF, also present in the normal ileum at the bottom of the crypt?

Mahida: We haven't looked at TNF. We have looked at mesenteric lymph node cells in terms of their Th1/Th2 profile. There is a Th1-type response early and a Th2 response at around day 8–10 of infection. We postulate that these lymph node cells are those that have recently migrated from the gut: we haven't looked directly at the lamina propria T cells.

Rhodes: The hypothesis could be that the normal anatomical localization of the Paneth cells at the bottom of the crypt could be driven by T cells.

Mahida: The other cell we postulate might be involved is the myofibroblast.

Jewell: Why do some people with chronic ulcerative colitis and in remission have a high density of Paneth cells, and in other people it is hard to find one?

Rhodes: Perhaps it could be TNFα polymorphism.

Mahida: I don't think this is a response specific to IBD, disappointing as that might be. There are some situations, for instance with inflammation in association with diverticular disease, where Paneth cell metaplasia also occurs. One of the reasons we were pursuing the parasite model was to try to get a handle on what is driving the stem cells in the colonic crypt to differentiate along the Paneth and goblet cell pathway.

Bjarnason: Being at risk of someone putting me on antidepressants, I wonder whether the Paneth cell is now the last cell in the normal gut that has up to now not been implicated in the aetiology of inflammatory bowel disease. Is there any cell you can think of that hasn't been?

Mahida: We may have a situation where one or two cell types may predispose one to developing disease, but subsequently all the other cells come into the equation, including T cells, myofibroblasts, macrophages and so on. The reason that Paneth cells are now of interest is twofold. First, there is the issue of potential novel antibiotics, in an era where we are getting short of antibiotics against bacteria that are resistant to multiple agents. The second issue is in understanding the pathogenesis of inflammatory bowel disease.

Sartor: Could we expand a little on the specificity of commensal bacteria? Is there any specificity among the isoforms of the β-defensins?

Mahida: The only data that I have seen have been in relation to *Staphylococcus aureus*. HBD1 and 2 have been reported to have limited activity whereas HBD3 and HBD4 express potent activity.

Sartor: I'd love to hear from other people. I hear that α-defensin deficiency may be relatively specific for small bowel Crohn's, and β-defensin deficiency for colonic Crohn's. And bacteroides will respond to defensins but not bifidobacteria. In broad terms, bifidobacteria are protective and bacteroides detrimental. Is this an explanation? The specificity of the α-defensins may give us a guide to antibiotics that may be relevant to small bowel Crohn's.

Stange: Absolutely. I think the key would be to find substances that induce defensin synthesis at the site where there is inflammation. I am rather sure that it will not be the solution to administer defensins orally. This could be pretty dangerous, because if you give natural antibiotics to a broad range of people, resistance could develop which might be devastating to the whole population. Natural antibiotics such as defensins have been working for tens of thousands of years, but if we drive resistance this could be a disaster. We are going the other way, looking for substances such as isoleucine that might drive defensin synthesis.

Sartor: I would argue that if there is a defective production of α-defensins in the inflammatory setting, perhaps they will not respond to exogenous drivers.

Stange: That is true for the normal signalling pathway. This is probably why *E. coli* doesn't work in Crohn's disease, because it uses the normal complex signalling pathway. But if you use a small molecule, it would be different. Decent-sized companies will have thousands of small compounds that they could analyse through high-throughput analysis, and perhaps come up with 100 effective compounds. 90 of these might be toxic but we could end up with a substance that is not toxic and induces defensins. The application could be much broader than just Crohn's disease. Look at chronic obstructive pulmonary disease (COPD), for example: there are recurrent infections which trigger disease progression. If you had a drug that pushed up defensin formation, this could have enormous impact.

Sartor: With regard to the specificity for α-defensins, is there anything known about their effect on commensal organisms?

Mahida: In *in vitro* studies, the precursor form of HD5 is active against *Listeria monocytogenes*. The mature form of HD5 is active against *Salmonella typhimurium*, *E. coli* and a few others. The full range of the activity hasn't been studied.

Sartor: To me this would seem like a very important area to study.

Mahida: Yes. We have looked at the activity on different bacterial strains and have done further work on antimicrobial activities released by cells that are undergoing apoptosis.

Furrie: We haven't looked at α-defensins. We are currently studying a wide range of commensals with the recombinant β-defensins, and we are collating the data. *Enterococcus faecalis* loses about a log number over 2 hours exposure but can survive with them. In the literature there does seem to be a Gram-positive/Gram-negative split, where the Gram-negatives are sensitive. But we find that some Gram-positives are quite sensitive to defensins. There isn't a simple division.

Sartor: Remember that there is a gradient, and you may have very high concentrations in the mucus biofilm.

Furrie: As soon as the defensins leave the producing epithelial cell they will be destroyed due to their sensitivity to salt. All antimicrobial action has to be very close to the epithelium. I suspect this is the whole point of them, because if they were floating about in the lumen you would completely skew the lumenal populations.

Sartor: This was my immediate feeling when you showed the sterile crypt.

Mahida: What has been suggested is that in the normal situation the Paneth cell-derived α-defensins may play a role in maintaining sterility of the crypt, thereby protecting the stem cells against microorganisms. However, we don't really know the concentrations of antimicrobial peptides present in the crypt lumen. Have you or Eduardo Stange done any work to see whether the epithelial secretory products, which may be β-defensins, have any activity against the resident flora?

Stange: We are in the process of doing this. We have developed a system where we can follow bacterial killing by FACS (fluorescence-activated cell sorting) analysis. This is nice because it is quantitative, we can use different bugs, and look at different defensins and extracts. If we go one step back and look at work published by Colombel's group (Darfeuille-Michaud et al 1998) on adherent *E. coli* specific for Crohn's in the terminal ileum, or Swidsinski's data (Swidsinski et al 2002) reporting a biofilm of mixed flora in the colon of UC and Crohn's and not in controls, I have always had a problem explaining this as a T cell problem. It is obviously an epithelial deficiency. The most logical thinking to me is that this represents a defensin deficiency. This is why you observe adherent bacteria, some of which may invade and cause inflammation. There also are a few patients who carry pathogens like mycobacteria or mycoplasma. Why is this? Because they have a problem in their antibiotic defence.

Sartor: The problem with the specificity is that UC patients also have increased adherent bacteria, and inflammatory controls do too.

Stange: I can't explain UC.

Rhodes: We have got some data using a similar technique to Swidsinski, and we have failed to find increased mucosa-associated *E. coli* under the mucus layer in UC, although we do find them in Crohn's. If you look at the Swidsinski paper carefully they only just reach significance in UC. This is curious: from the previous microscopic studies from Adrian Allan's group the mucus layer is thicker in Crohn's. But for some reason there are more bugs under the mucus layer in Crohn's.

Stange: This fits with what we see, which is up-regulated defensin formation in UC.

Fedorak: Can you induce bacterial resistance to these defensins? Is this not a problem with inadequate defensin production? The bacterial population could have mutated so that they no longer respond to the defensins.

Stange: There are bugs that are resistant and some actually down-regulate defensin production, like *Shigella*. This is a survival mechanism for *Shigella*.

Mahida: Shigella seems to have an interesting virulence mechanism in terms of reducing HBD1 and LL37 expression.

Stange: One bug which has developed resistance is *Staphylococcus*. This has developed resistance against defensins formed by the skin. What is the typical infection if you have a wound? It is *Staphylococcus*. This is because of this break in the antibiotic resistance. It is a constant war.

Gibson: I'd like to go a bit further with the hypothesis that Crohn's disease is due to α-defensin deficiency. How do you explain the fact that it is a western disease that is now showing an increased prevalence in Japan?

Stange: The idea is basically that if you grow up in a dirty environment you learn how to induce defensins. As a worst case scenario, if a child is unable to induce its defensins in a third world environment, it will die from infections. The major cause of child mortality in the third world is severe diarrhoea. We are growing up in such a clean environment that we 'forgot' how to produce defensins. There is also the parasite issue. If Professor Mahida's data can be translated into human physiology, this would mean that if you have a parasite environment there are high levels of defensins all the time, and then you are resistant to developing Crohn's and UC.

Rhodes: If defensin formation is TNF mediated then infliximab will wipe out the defensins.

Stange: Some infliximab patients develop sepsis and die (Andus et al 2003). It could be bacterial translocation.

References

Andus T, Stange EF, Hoffler D et al 2003 Suspected cases of severe side effects after infliximab (Remicade) in Germany. Med Klin 15:429–436

Creamer B, Pink IJ 1967 Paneth-cell deficiency. Lancet 1:304–306

Darfeuille-Michaud A, Neut C, Barnich N et al 1998 Presence of adherent *Escherichia coli* strains in ileal mucosa patients with Crohn's disease. Gastroenterology 115:1405–1413

Fellermann K, Wehkamp J, Herrlinger KR et al 2003 Crohn's disease: a defensin deficiency syndrome? Eur J Gastroenterol Hepatol 15:627–634

Ghosh D, Porter E, Shen B et al 2002 Paneth cell trypsin is the processing enzyme for human defensin-5. Nat Immunol 3:583–590

Hooper LV, Stappenbeck TS, Hong CV, Gordon JI 2003 Angiogenins: a new class of microbicidal proteins involved in innate immunity. Nat Immunol 4:269–273

Ouellette AJ, Selsted ME 1996 Paneth cell defensins: endogenous peptide components of intestinal host defense. FASEB J 10:1280–1289

Salzman NH, Ghosh D, Huttner KM, Paterson Y, Bevins CL 2003 Protection against enteric salmonellosis in transgenic mice expressing a human intestinal defensin. Nature 422:522–526

Swidsinski A, Ladhoff A, Pernthaler A et al 2002 Mucosal flora in inflammatory bowel disease. Gastroenterology 122:44–54

Intestinal mycoplasma in Crohn's disease

W. E. W. Roediger

Department of Surgery, University of Adelaide, The Queen Elizabeth Hospital, Adelaide, SA 5011, Australia

Abstract. Intestinal diversion with reconnection in active Crohn's disease (CD) indicates that luminal contents or bacteria contribute to the formation of CD lesions. Fluorescent staining for mycoplasma in freshly resected Crohn's tissue and electron microscopy reveal intracellular organisms akin to mycoplasma. Historically, tissue culture of CD has shown mycoplasma described as contaminants. Mycoplasma are surface epithelial parasites requiring exogenous cholesterol for membrane stability and cell entry. PCR of intestinal tissue has shown *Mycoplasma pneumoniae* to be detectable more significantly in CD. Oral *M. iowae* in experimental poultry localizes to the distal small bowel and colon. Hypothetically, lipopeptides of mycoplasmal membranes are proposed to cause chronicity and stronger immune responses than by other bacteria. 'Intestinal' mycoplasmas, from a number of observations, deserve consideration as organisms mediating inflammation of acute and chronic CD.

2004 Inflammatory bowel disease — crossroads of microbes, epithelium and immune systems. Wiley, Chichester (Novartis Foundation Symposium 263) p 85–98

Mycoplasma of the intestinal tract have not been subjected to systematic study even though several species have been reported from the gut (Roediger & Macfarlane 2002) and oral cavity of humans (Tully 1993). The hypothesis that intestinal mycoplasma might be responsible for Crohn's disease (CD) was first proposed in 1991 (Roediger 1991), preceded by reports that L forms of bacteria (Parent & Mitchell 1978) and subspecies of *Mycobacteria*, foremostly *Mycobacteria avium paratuberculosis* (Map), might be causative of Crohn's disease. The involvement of Map in CD is contentious (Collins 2002, Bölske et al 2003) mainly because acid-fast bacilli, *in situ* hybridization and fluorescent antibody staining for these organisms have been negative in Crohn's tissue. Infestation of the gut mucosa by *M. tuberculosis* and *M. bovis*, however, has indicated that several species of mycobacteria may lead to intestinal disease but not of the Crohn's type. That mycoplasma species can cause intestinal disease can be adduced from several lines of evidence (Table 1).

TABLE 1 Factors in consideration of mycoplasmal involvement in Crohn's disease

Fluorescent cytoplasmic DNA staining
Electron microscopy
Enteral nutrition and lipid content of diet
Clinical associations
Evidence from tissue culture
Antibiotic trials
Polymerase chain reaction for mycoplasma
Animal sources of mycoplasma
Lipopeptides (MALP) of mycoplasmal origin
Mycoplasmal immune cell responses

Mycoplasmal attributes

Approximately 110 species of mycoplasma are known (Madigan et al 2000) all noted for their small genome size. About 20 genes in mycoplasma are devoted to bioenergetics compared to ±150 or more in bacteria such as *Escherichia coli* (Castresana 2001). Mycoplasma have no conventional cell membrane but a limiting membrane with which cholesterol is secondarily associated. Exogenous cholesterol is needed for their growth as none is synthesized by the microorganism. The mycoplasmal membrane has a unique lipoprotein or lipopeptide structure which is distinct from that found in other bacteria (Razin et al 1998). Mycoplasmas have a distinct interaction with the immune system and are mitogenic for lymphocytes. Membrane properties permit mycoplasma to undergo cell fusion with eukaryote host cell membranes. Partly for this reason they have been referred to as 'epithelial parasites'. The membrane properties of mycoplasma do not permit dye uptake, making staining for light microscopy impossible.

Mycoplasmal DNA staining

Because chemical characteristics of the limiting membrane of mycoplasma make staining for light microscopy impossible, quality control for mycoplasmal contamination of cell cultures has rested on DNA fluorescent staining with Hoechst 33258 or H stain. This is a nucleic acid groove binding dye that fluoresces and has specificity for AT-rich regions (Teng et al 1988). Mycoplasma-contaminated cultures show fluorescent bodies surrounding host cell nuclei and the dye tested positive with at least 17 mycoplasmal species with which cells were infected (Chen 1977, Hay et al 1989). The method of Chen (1977) was adapted for cryostat sections of ileal mucosa adjacent to Crohn's ulcers and positive cytoplasmic granules found as in tissue cultures (Fig. 1). Both mitochondrial nucleic acid staining and other bacterial staining could be excluded on the basis of size and distribution. The presence of intracytoplasmic fluorescence resembling mycoplasma was patchily aggregated in separated groups of crypt cells.

Molecular localization of mycoplasma

Genomic localization of mycoplasma has been demonstrated by means of *in situ* hybridization (Boye et al 2001) and polymerase chain reaction (PCR) (Uphoff & Drexler 2002) of specific nucleic acid sequences of mycoplasma. Both procedures require the use of specific oligonucleotides or probes targeted at sequences that might be either species or genus specific. Compared with tissue culture for mycoplasma, *in situ* hybridization is very accurate but no studies with this technique have been carried out in Crohn's disease.

A study using PCR to look for *Mycoplasma pneumoniae* in diseased and healthy intestinal mucosa was recently reported (Chen et al 2001). Three biopsies of each case were tested. The primers used were 345 bp portion of the PI adhesin gene of *M. pneumoniae* and products identified by electrophoresis or ELISA. DNA was significantly higher in biopsies from patients with CD (59.2%) than from those with ulcerative colitis (UC) (26.3%) or non-inflammatory bowel disease (IBD) controls (37.7%). While the above is a species-specific analysis of mycoplasmal oligonucleotides, genus-specific oligonucleotides are also available (van Kuppeveld et al 1992) named MGSO and GPO-1 of the V2 and V3 region according to the nomenclature of Neef. By using such primers on single biopsies, we were able, using gel electrophoresis, to detect mycoplasma in three out of six acute Crohn's colitis but none of three chronic stenosing CD. *M. pneumoniae* was used as an internal standard (Fig. 2). These observations suggest that at least 3–4 biopsies per case are needed and that detection would be optimal in acute ulceration rather than in chronic fibrosing disease. Site-specific sensitivity of mycoplasma detection by PCR has been found in animals (Kurth et al 2002). A further lesson of PCR detection of microorganisms is that of Map detection where the IS900 probe initially yielded strong positive results between 1992–1996 followed by subsequent less positive reports (Quirke 2001). The specificity of the IS900 oligonucleotide for Map has been called into question (Quirke 2001, Bölske et al 2003) which should also be a consideration for the probes used in CD. PCR results in CD have to be interpreted with care and emphasis placed on specificity and pathogenicity. Future analyses hopefully will provide answers.

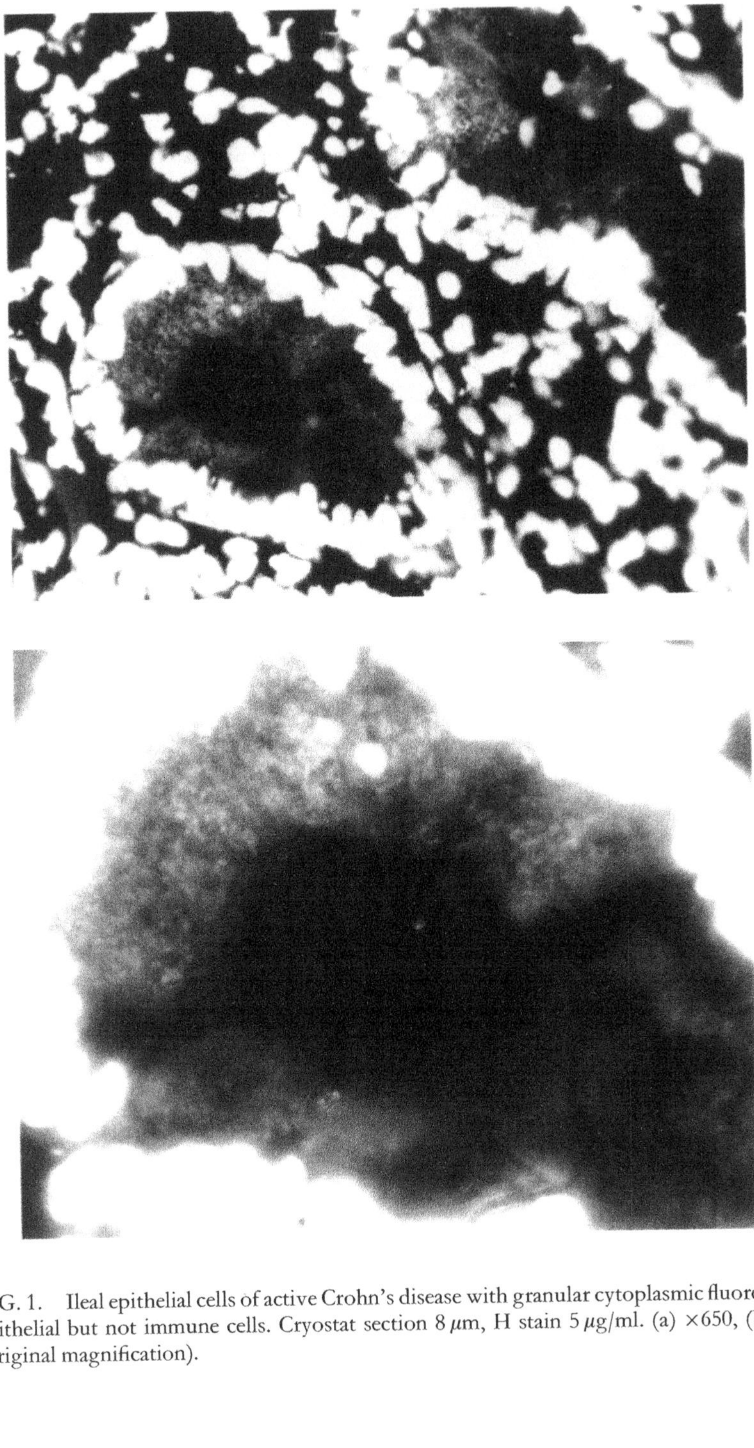

FIG. 1. Ileal epithelial cells of active Crohn's disease with granular cytoplasmic fluorescence in epithelial but not immune cells. Cryostat section 8 μm, H stain 5 μg/ml. (a) ×650, (b) ×1000 (original magnification).

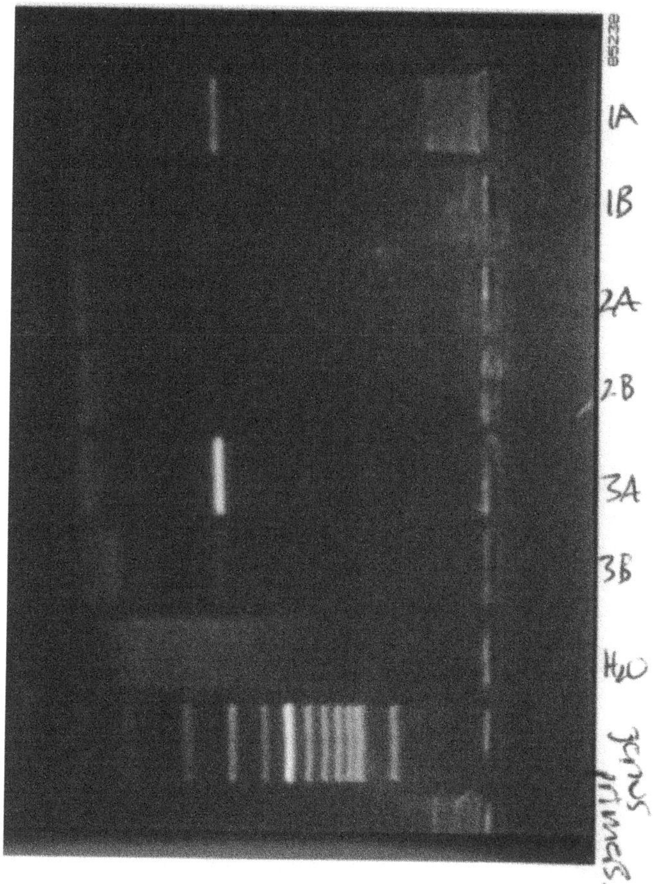

FIG. 2. Gel electrophoresis of PCR *Mycoplasma* (genus specific) product of two patients (1A, B, 2A, B) with controls of *M. pneumoniae* (3A, B). A, B at two dilutions with positive reaction in 1A and B.

Animal sources of mycoplasma

Gastrointestinal diseases such as *Mycobacterium bovis* enteritis and *Campylobacter jejuni* enteritis are primarily transmitted by bacteria from animal sources, that is cattle and poultry. The main sources of mycoplasma in nature are plants and animals, of which plant species have not been recorded in humans. Poultry are affected by mycoplasma infections and also carry mycoplasma as commensals (Calnek 1997). The first historical assignment of disease to mycoplasma was entitled 'Epizootic pneumo-enteritis of the turkey' (Dodd 1905). The term 'enteritis' has since been dropped as the most noticeable site of infection in

TABLE 2 Factors related to the chronicity of mycoplasmal infection

Intracellular DNA replication
Lipopeptide structure and lipid composition (also known as macrophage activating lipopeptide, MALP)
Recognition or interaction of TOLL-like receptor with lipopeptide
NOD2 cytoplasmic protein related to TOLL-like signalling pathway

poultry has been the respiratory tract. Avian mycoplasmas have been extensively studied and certain observations are of relevance to CD. Oral challenge of turkey poults with *Mycoplasma iowae* led to the most frequent recovery from the mucosa of the distal portion of the small intestine, caecum and large intestine suggesting that recovery of *M. iowae* from these tissues indicated active infection (Shah-Majid & Rosendal 1987). While Crohn's-like lesions were not detected in the avian gut the siting within the gastro-intestinal tract may be of importance. Studies in turkey embryos also led to the conclusion that *M. iowae* had a predilection for the intestinal tract of avian hosts (Mirsalimi et al 1989). Mycoplasma infection in turkeys and pigs (Hagedorn-Olsen 1999) are known to cause arthritis, a clinical feature not infrequently found in CD.

Chronicity of CD and lipopeptides

A key feature of CD is chronicity of inflammation that is attenuated but not eradicated by immune suppression. Mycoplasma infection in humans may pursue a chronic course for which several factors can be invoked, as outlined in Table 2.

Adhesion of mycoplasma to cells is well established but intracellular location of viable mycoplasma is sometimes denied yet known from tissue culture and mycoplasma infections of AIDS subjects. Experiments have shown that intracellularly localized mycoplasma maintained infectivity of tissue for more than 6 months (Dallo & Baseman 2000). The conclusions were that pathogenic mycoplasma reside and replicate intracellularly over extended periods in human cells consistent with the ability of mycoplasmas to circumvent antibiotic therapy and immune surveillance to establish chronic infection.

Stimulation of macrophages and release of cytokines are produced by microbial cell wall products such as lipopolysaccharides, lipoteichoic acid or lipopeptides (lipoprotein), the latter best characterized in mycoplasma. Membrane proteins of mycoplasma were first shown to release cytokines in 1994. The chemical characteristics of mycoplasmal lipopeptide were then established and synthetic derivatives of the mycoplasmal membrane product prepared (Mühlradt et al

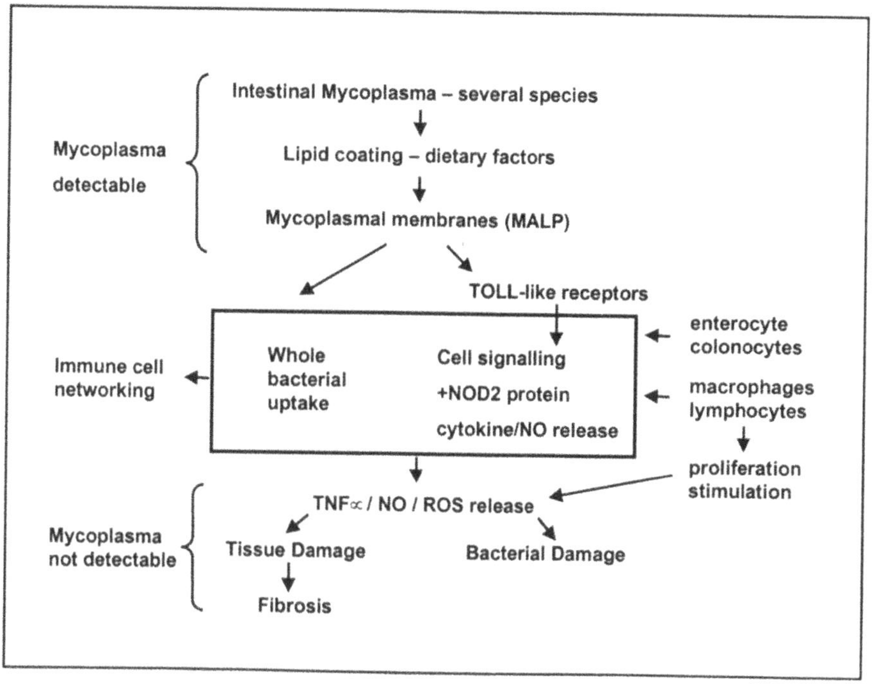

FIG. 3. A model of intestinal mycoplasma in the pathogenesis of Crohn's disease.

1997) which had actions equivalent to the natural mycoplasmal lipopeptide. The synthetic mycoplasmal product has been termed macrophage-activating lipopeptide (MALP) and found to be 50–100 times more active, weight for weight, than lipopolysaccharide from *E. coli* (Mühlradt et al 1998). MALP induces proliferation of lymphocytes (Weigt et al 2003) and reproduces inflammatory changes in lungs (Lührmann et al 2002). We are currently examining the ability of MALP to induce inflammation in the mucosa of small and large intestines of experimental animals.

The key mediators of microbe detection by cells are TOLL-like receptors (TLRs) (Underhill & Ozinsky 2002) sited across immune and epithelial cell membranes with an extracellular arm and a cytoplasmic domain that triggers a pathway of cell signals leading to cytokine release (Yeh & Chen 2003). Signals of TLRs can be amplified or suppressed by cytosolic proteins termed NOD (Inohara et al 2002) and these proteins can negate disease resistance by yet unknown mechanisms. The lipopeptides of mycoplasma are recognized by TLRs (Morr et al 2002) and such recognition is dependent on the structure of the lipopeptide.

Chronicity of mycoplasma infection could be determined by (1) type and density of luminal mycoplasma, (2) lipopeptide structure, (3) presence and type of TLRs and (4) degree and action of NOD proteins. Further investigations in Crohn's disease in consideration of all these points are warranted.

A model of the pathogenesis of Crohn's disease

Drawing together information from evidence previously collated (Roediger 1991, Roediger & Mcfarlane 2002) and now newly presented, it has become possible to draw up a model of the pathogenesis of Crohn's disease (Fig. 3). The human body has been used as a microbial observatory and there may well be a thin line between commensal and pathogen (Gilmore & Ferretti 2003), which requires careful scrutiny. With appropriate reconsideration of Koch's postulates in the genomic era (Fredericks & Relman 1996) the steps of causation in Crohn's disease may hopefully be untangled in the next decade.

References

Bölske G, Englund S, Johansson K-E, Königsson MG 2002 False positive *Mycobacterium avium* subsp. *paratuberculosis* IS*900* PCR and its diagnostic implications. Proceedings of the Seventh International Colloquium on Paratuberculosis 7:261–266

Boye M, Jensen TK, Ahrens P, Hagedorn-Olsen T, Friis NF 2001 In situ hybridisation for identification and differentiation of *Mycoplasma hyopneumoniae*, *Mycoplasma hyosynoviae* and *Mycoplasma hyorhinis* in formalin-fixed porcine tissue sections. APMIS 109:656–664

Calnek BW (ed) 1997 Diseases of poultry. 10th edn. Iowa State University Press, Ames, Iowa, USA

Castresana J 2001 Comparative genomics and bioenergetics. Biochim Biophys Acta 1506: 147–162

Chen TR 1977 In situ detection of mycoplasma contamination in cell cultures by fluorescent Hoechst 33258 stain. Exp Cell Res 104:255–262

Chen W, Li D, Paulus B, Wilson I, Chadwick VS 2001 High prevalence of Mycoplasma pneumoniae in intestinal mucosa biopsies from patients with inflammatory bowel disease and controls. Dig Dis Sci 46:2529–2535

Collins MT 2002 Summation. Proceedings of the Seventh International Colloquium on Paratuberculosis 7:531–535

Dallo SF, Baseman JB 2000 Intracellular DNA replication and long-term survival of pathogenic mycoplasmas. Microb Pathog 29:301–309

Dodd S 1905 Epizootic pneumo-enteritis of the turkey. J Comp Path Therapeutics 18:239–245

Fredericks DN, Relman DA 1996 Sequence-based identification of microbial pathogens: a reconsideration of Koch's postulates. Clin Microbiol Rev 9:18–33

Gilmore MS, Ferretti JJ 2003 The thin line between gut commensal and pathogen. Science 299:1999–2002

Hagedorn-Olsen T, Nielsen NC, Friis NF 1999 Induction of arthritis with *Mycoplasma hyosynoviae* in pigs: clinical response and re-isolation of the organism from body fluids and organs. Zentralbl Veterinarmed A 46:317–325

Hay RJ, Macy ML, Chen TR 1989 Mycoplasma infection of cultured cells. Nature 339:487–488

Inohara N, Ogura Y, Nuñez G 2002 Nods: a family of cytosolic proteins that regulate the host response to pathogens. Curr Opin Microbiol 5:76–80
Kurth KT, Hsu T, Snook ER, Thacker EL, Thacker BJ, Minion FC 2002 Use of a *Mycoplasma hyopneumoniae* nested polymerase chain reaction test to determine the optimal sampling sites in swine. J Vet Diagn Invest 14:463–469
Lührmann A, Tschernig T, Pabst R 2002 Stimulation of bronchus-associated lymphoid tissue in rats by repeated inhalation of aerosolised lipopeptide MALP-2. Pathobiology 70:266–269
Madigan MT, Martinko JM, Parker J 2000 Biology of microorganisms. 9th edn. Prentice Hall International Inc, New Jersey
Mirsalimi SM, Rosendal S, Julian RJ 1989 Colonization of the intestine of turkey embryos exposed to Mycoplasma iowae. Avian Dis 33:310–315
Morr M, Takeuchi O, Akira S, Simon MM, Mühlradt PF 2002 Differential recognition of structural details of bacterial lipopeptides by toll-like receptors. Eur J Immunol 32:3337–3347
Mühlradt PF, Kiess M, Meyer H, Süssmuth R, Jüng G 1997 Isolation, structure elucidation and synthesis of a macrophage stimulatory lipopeptide from *Mycoplasma fermentans* acting at picomolar concentration. J Exp Med 185:1951–1958
Mühlradt PF, Kiess M, Meyer H, Süssmuth R, Jung G 1998 Structure and specific activity of macrophage-stimulating lipopeptides from *Mycoplasma hyorhinis*. Infect Immun 66:4808–4810
Parent K, Mitchell P 1978 Cell wall defective variants of pseudomonas-like (Group Va) bacteria in Crohn's disease. Gastroenterology 75:368
Quirke P 2001 Antagonist. *Mycobacterium avium* subspecies paratuberculosis is a cause of Crohn's disease. Gut 49:757–760
Razin S, Yogev D, Naot Y 1998 Molecular biology and pathogenicity of Mycoplasmas. Microbiol Mol Biol Rev 62:1094–1156
Roediger WEW 1991 A new hypothesis for the aetiology of Crohn's disease—evidence from lipid metabolism and intestinal tuberculosis. Postgrad Med 67:666–671
Roediger WEW, Macfarlane GT 2002 A role for intestinal mycoplasmas in the aetiology of Crohn's disease? J Appl Microbiol 92:377–381
Shah-Majid M, Rosendal S 1987 Oral challenge of turkey poults with Mycoplasma iowae. Avian Dis 31:365–369
Teng M-K, Usman N, Frederick CA Wang A H-J 1988 The molecular structure of the complex of Hoechst 33258 and the DNA dodecamer d(CGCGAATTCGCG). Nucleic Acids Res 16:2671–2690
Tully JG 1993 Current status of the mollicute flora of humans. Clin Infect Dis 17(suppl 1):S2–9
Underhill DM, Ozinsky A 2002 Toll-like receptors: key mediators of microbe detection. Curr Opin Immunol 14:103–110
Uphoff CC, Drexler HG 2002 Comparative PCR analysis for detection of mycoplasma infections in continuous cell lines. In Vitro Cell Dev Biol Anim 38:79–85
Van Kuppeveld FJM, Van Der Logt JTM, Angulo AF et al 1992 Genus- and species-specific identification of mycoplasmas by 16S rRNA amplification. Appl Env Microbiol 58: 2606–2615
Weigt H, Mühlradt PF, Emmendörffer A, Krug N, Braun A 2003 Synthetic mycoplasma-derived lipopeptide MALP-2 induces maturation and function of dendritic cells. Immunobiol 207:223–233
Yeh W-C, Chen N-J 2003 Immunology: another toll road. Nature 424:736–737

DISCUSSION

Sartor: This is extremely provocative, and once again you have broadened our horizons. These results are very preliminary, of course. It is an attractive hypothesis because mycoplasma are present in the food chain. I recall an extremely interesting study done in British children, in which they looked at serologic responses and tried to associate infections with flares of clinically apparent IBD. 50–60% of flares were associated with serological responses and *M. pneumoniae* was number one on the list. The concerns are that if you have an open ulcer and something in the food chain, could this be just secondary uptake. The studies used non-specific DNA. There could be other bacteria there. The intracellular organelles you showed are reminiscent of the organelles shown in electron microscopy studies. It would be interesting to go back and use *in situ* hybridization and immunoelectron microscopy with your 16S ribosomal DNA probe. I couldn't tell from your EM studies whether this was epithelial versus lamina propria. Was this associated with an ulcer or intact epithelium? The other question that needs to be addressed is are the serologic and T cell responses to specific antigen in your Crohn's patients?

Roediger: You have raised a lot of points. Serology is difficult. We looked in quite a few of our patients for *M. pneumoniae* because the serotype is available. But *M. pneumoniae* can change its serotype, and there is no cross-reaction with other mycoplasma, so we might have a mycoplasma infection by another species. The poultry industry has developed a whole battery of possibilities of measuring various serotypes of mycoplasma. I think we should transfer this knowledge to the clinical setting. To answer your other question, those pictures of epithelial cells were at the edge of an ulcer. We thought this would the best place to look. I haven't looked at the lamina propria. As you know, this is a highly controversial field. There are mycoplasma infections in AIDS and arthritis, and I should add that a lot of the mycoplasma organisms do cause arthritis. In CD 25% of patients have arthritis. There is no other organism that infects both intestinal tract and produces arthritis.

Sartor: You have your primers in a broad area. You could find out in your CD positives which species are there and you could then selectively search by serology.

Roediger: Yes, indeed.

Sartor: Is erythromycin an effective antibiotic?

Roediger: I use clarithromycin clinically because this is where the work has been reported. I'm not sure about the cross-over from erythromycin to clarithromycin: in theory it should work.

Jewell: Back in the early 1980s, a surgical trainee called Peter Harper and I took a group of patients with Crohn's colitis who had had the colon defunctioned with a split ileostomy. We took ileostomy contents, dripped them into the colon and produced an inflammatory reaction. What Peter then did was to centrifuge ileostomy contents and pass them through a 22 nm filter. The ultrafiltrate was then put into the distal colon and there was no reaction. Would a mycoplasma go through a 22 nm filter?

Roediger: No, it would be filtered out. With regard to instillation into the defunctioned colon, we have to be mindful of diversion colitis. The epithelium is different in the defunctioned colon.

Jewell: I thought there was a little disparity between the initial DNA stains that you showed where there was very heavy staining, and the EM where you had to look quite hard to find these structures. If PCR is so sensitive one might have expected you to have picked up a greater percentage of patients.

Roediger: With the genus-specific PCR that should be achievable. We didn't want to repeat the PCR for *M. pneumoniae* which was done in New Zealand (Chen et al 2001). The staining preceded this work, and the technology has been evolving (Chen 1977). We are keen that someone should take this on in a systematic way, but we haven't got around to this. It needs some determination. We need to cut a lot of cryostat sections and look at them systematically. If you just look in isolated sections it ends up unidentifiable.

Rhodes: I have a comment about mycobacterial immunoreactivity. This is an intriguing area. We know that about two-thirds of patients with CD are anti-*Saccharomyces cerevisiae* positive. The epitope for this antibody is a mannan with a specific mannose-1,3-mannose linkage which is not necessarily specific to *S. cerevisiae*. Two papers (Michell et al 2003, Guerardel et al 2002) have appeared over the last 12 months from different groups, neither of them researching CD, looking at cell wall glycoproteins in mycobacteria. It turns out that some mycobacterial species have exactly the same 1,3-mannose linkage on the cell surface. One of these papers (Michell et al 2003) has been looking at the BCG variant of *Mycobacterium bovis* which not only expresses this mannose linkage, but also they have shown that the presence of this carbohydrate structure is critical to the immunogenicity of the BCG vaccination. It goes down 90% if the carbohydrate is removed. They have also shown that *E. coli* doesn't express the mannosyl transferase necessary to give this structure. Frustratingly, from our point of view, they haven't looked at *Mycobacterium paratuberculosis*. This is compatible with the idea that the ASCA reactivity might be driven by very low level mycobacterium infection. The question I would like to ask you is related to antibiotics, and mycoplasma. The thing that puts me off the mycoplasma idea is the lack of permanent response to antibiotics, which one thought might have eradicated it. We use clarithromycin a lot: we use it because it gets high penetration inside cells. It is probably going to kill quite a lot of different bacteria. We see very good short-term responses, but what we don't get is long-term cure. It will be intriguing to see whether your Australian collaborative antimycobacterial trial gives long term cure. I don't think John Herman-Taylor has convincingly reported long-term cure, either.

Roediger: With bacteria and bacterial fragments, the response to antibiotics is different. With mycoplasma infection the whole viable organism is not needed:

mycoplasma membranes alone in macrophages will give an immune response that lasts for 6–12 months.

Rioux: There have been many studies on the genetics and immunology of mycobacterial infection in mice. Is there anything we can learn from those studies? Do they have any gastrointestinal effect?

Roediger: I am not aware of this.

Sartor: In general, with parasites the genetic background of the mouse and its propensity to form a Th1 versus Th2 immune activation will very much determine whether the animal dies or clears the infection. Surely genetics must play a role.

Rioux: Was there any gastrointestinal inflammation in those models?

Roediger: Not as far as I know.

Parkos: Two quick asides. First, it is interesting that the MALP2 protein not only has potent effects on macrophages, but also on neutrophils. Second, since we do almost all *in vitro* work, mycoplasma infection of cell lines is a huge problem. We do PCR on everything and treat infected cells with the appropriate drugs, but often they become infected again. We have always attributed this to user contamination, for example, but it is an ongoing issue.

Roediger: Novartis have an antibiotic that is used in the poultry industry, which they suggest clears animals of mycoplasma.

Parkos: We have tried two or three antibiotics. We can clear mycoplasma so it is not detectable by PCR. Personally, I think that if you are looking at the Caco-2 cell literature, with some reports, if authors aren't maintaining that their cell lines are mycoplasma free, then I question whether some responses are mycoplasma related.

Kelleher: It is possible to eradicate mycoplasma from cell cultures as long as you transfer them from the environment in which they are currently contained. Clearance is feasible.

Wilkins: You still have to wonder whether it might not be an opportunistic commensal that increases in disease conditions. You have mentioned that chickens don't get IBD when exposed to mycoplasma. Has anyone tried introducing mycoplasma to germ-free mice?

Roediger: I am not aware of any studies in germ-free mice but you raise a very important point concerning the transition from commensal to pathogenic organism. There is a question mark in the model I put up. I assumed mycoplasma are commensals, but I am not sure what is required for the transfer into the pathogenic form.

Meddings: I was interested that mycoplasma tend to invade lower down in the gut, and that they also suck up cholesterol and other lipoproteins. There seems to be a discordance here because the concentrations of these are highest in the proximal gut. Is this an important invasive mechanism?

Roediger: The membrane stability is maintained by sterols. Whether this then becomes a disguise for the enterocyte to presume it is a micelle I am not sure. It is all a bit of a black box.

Schölmerich: There are a number of plant sterols that go further down in the gut. There might be enough around.

Bill Roediger, did you look at any other inflammation, such as diverticulitis? Have you seen anything similar?

Roediger: Diverticular colitis is an entity that is becoming more recognized, and we have looked there. We find macrophages and signs of bacteria but have only done one case. It depends on how intensely one looks.

Gibson: In tissue culture, the presence of mycoplasma is an almost normal event. The problem Jon Rhodes was alluding to was why do you not get cure after triple therapy for paratuberculosis? This is a critical question. The fact is we are all exposed to *M. paratuberculosis* on a regular basis. These issues with mycoplasma and mycobacteria might not be one of exposure: it might be just one of host response to the organisms. A critical theme is going to be determining whether the mycoplasma you find are any different. They may be the same, and it may be the macrophage response to the mycoplasma that is causing the problem. This could be why chronic therapy with erythromycin has to be given, rather than short-term therapy. In the paratuberculosis study in Australia, we have seen patients who have had prolonged remission. This double-blind study has yet to be unblinded, but it is hard to believe that they could have a prolonged remission on placebo. We had two patients with brilliant prolonged remission who have both since relapsed. It must be re-exposure.

Sartor: Peter, could you comment on the status of your multicentre study?

Gibson: The recruitment is finished. The patients have all finished the treatment arm and are now in the one year follow-up arm, which should finish in September 2004, with the results available mid-2005.

Jewell: Of course, response to that doesn't say anything about the aetiology of disease, because you are using such a broad spectrum of antibiotics.

Gibson: Do we know anything about mycoplasma resistance to antibiotics: do they develop resistance?

Roediger: Once the mycoplasma are intracellular, they cannot be attacked by antibiotics. One can't eradicate intracellular mycoplasma. This is a problem with this organism. One has to almost rethink everything. The mycoplasma probably depend on toll-like receptors, which are a key element. There are lots of avenues to be explored.

Schölmerich: How can we treat *Mycoplasma pneumoniae* so well, then? It goes relatively fast and the disease doesn't recur.

Roediger: It is a prolonged treatment with a high recurrence rate.

Kelleher: Erythromycin is a bacteristatic antibiotic: it is not bactericidal.

Schölmerich: It works.

Sartor: If you prevent replication and transfer to another cell, and then that cell dies, then it is gone. How long does it persist within a cell?

Jewell: I have one patient who I look after jointly with a clinical immunologist, with common variable immunodeficiency. She has CD in the colon by any other criteria, but she also has intermittent problems with mycoplasma infection in the bladder. The mycoplasma infection in the bladder and the CD appear to behave totally independently. It is an interesting clinical scenario.

Rhodes: You made an intriguing remark about fat wrapping. I hadn't heard before that this definitely doesn't occur in tuberculosis. Can you comment on any sort of idea about pathogenesis for fat wrapping in CD? I gather that Jean-Frederic Colombel has pictures of magnetic resonance imaging (MRI) scans of children with CD where the fat wrapping on the MRI scan has become apparent before any barium abnormalities.

Roediger: Fat wrapping must be non-specific because in diverticulits an enormous amount of fat wrapping is seen. I interpret it to be part of the inflammatory component, although I am not sure how it is brought about.

Rhodes: So it definitely doesn't happen in intestinal tuberculosis (TB).

Roediger: I've only seen the inside of the peritoneal cavity in TB four times. The fat disappears; there is no fat wrapping.

Jewell: Bryan Warren is adamant that it doesn't occur in TB. The surgeons I've spoken to say they don't see fat wrapping in TB. Jean-Frederic Colombel thinks that tumour necrosis factor (TNF) is probably the trigger for fat wrapping. I can't believe that there isn't TNF in the inflammation caused by TB.

Rhodes: Colombel's idea was partly the other way round: the pro-inflammatory effect of the fat was causing the gut inflammation.

Sartor: He is saying that adipose cells actually produce the TNFα. I would point out that indomethicin-treated Lewis rats also have fat wrapping.

References

Chen TR 1977 In situ detection of mycoplasma contamination in cell cultures by fluorescent Hoechst 33258 stain. Exp Cell Res 104:255–262

Chen W, Li D, Paulus B, Wilson I, Chadwick VS 2001 High prevalence of Mycoplasma pneumoniae in intestinal mucosa biopsies from patients with inflammatory bowel disease and controls. Dig Dis Sci 46:2529–2535

Guerardel Y, Maes E, Elass E et al 2002 Structural study of lipomannan and lipoarabinomannan from *Mycobacterium chelonae*. Presence of unusual components with alpha 1,3-mannopyranose side chains. J Biol Chem 277:30635–30648

Michell SL, Whelan AO, Wheeler PR et al 2003 The MPB83 antigen from Mycobacterium bovis contains O-linked mannose and (1→3)-mannobiose moieties. J Biol Chem 278: 16423–16432

Probiotics, prebiotics and antibiotics in inflammatory bowel disease

John H. Cummings and San Choon Kong

Division of Pathology and Neuroscience, University of Dundee, Ninewells Hospital and Medical School, Dundee DD1 9SY, UK

Abstract. Probiotics and prebiotics are the sort of therapy that most patients with inflammatory bowel disease (IBD) wish for. They are without significant side effects, except possibly an increase in gas with prebiotics, and so far, appear to be entirely safe. However, are they effective? More than a dozen studies using probiotics in IBD have now been reported, and there is clear benefit in pouchitis and possibly also in Crohn's, although there are so many clinical types of this condition that a clear indication has yet to emerge. For ulcerative colitis (UC) more studies are needed. The use of prebiotics in IBD is only just starting, although significant effects on both the luminal and mucosa-associated flora have been demonstrated in healthy subjects. Antibiotics offer more certain hope in IBD treatment, although with a much greater risk of unwanted effects. Their efficacy in clinical studies varies, with Crohn's disease and pouchitis reporting more benefit than ulcerative colitis. However, the ideal combination of antibiotics, and rationale for their use has not been determined.

2004 Inflammatory bowel disease — crossroads of microbes, epithelium and immune systems. Wiley, Chichester (Novartis Foundation Symposium 263) p 99–114

Probiotics

Probiotics are viable bacteria, which are either added to food (usually dairy products) or taken as freeze dried supplements, that benefit health. From being considered a rather doubtful dietary constituent on the edge of normal nutrition and medicine, probiotics have come to occupy centre stage as evidence has accumulated that the normal commensal flora of the gut is important in both digestion and the body's defences, and can be modified by probiotics.

The commonly used probiotic bacteria belong usually to the bifidobacteria and lactobacillus genera. Some other species have been used including *Escherichia coli* Nissle and the yeast *Saccharomyces boulardii*. Research has now advanced to the point that probiotics have been shown to have a clear benefit (Table 1) in both preventing and reducing the severity of attacks of acute diarrhoea in children (Szajewska & Mrukowicz 2001, Van Niel et al 2002) in preventing antibiotic-associated diarrhoea (Cremonini et al 2002, D'Souza et al 2002), reducing the risk of atopy

TABLE 1 Probable health benefits of probiotics

Stimulation of immune function
Strengthen gut defence against pathogens
Reduction of atopy in children
Amelioration of acute diarrhoea principally (rotavirus in children)
Prevention of antibiotic associated diarrhoea
Treatment of inflammatory bowel disease
Treatment of pouchitis
Treatment of lactose intolerance
Prevention of genitourinary infections
Anti cancer effects (in animal models)

in children (Kalliomaki et al 2003) and in the treatment of IBD (Hart & Kamm 2003, Jonkers & Stockbrugger 2003). Other effects are summarised in the increasing numbers of reviews on the subject (de Roos & Katan 2000, Marteau 2002, Stephen et al 2002, Reid et al 2003).

Most of the clinical studies of probiotics have been largely empirical in their choice of which bacterial species to use, and little is presently understood of the mechanisms involved. However, over a remarkably wide range of physiological processes, probiotics exert an effect that is clearly beneficial to health. It is likely that probiotics will eventually find a place in the routine management of IBD, although there is a need for good clinical trials and further work on the interaction between probiotics, the commensal flora and mucosal function. Why then should we be interested in prebiotics?

Prebiotics

Prebiotics are non-digestible carbohydrates that selectively stimulate the growth of lactobacilli and bifidobacteria with benefit to health. Theoretically prebiotics should therefore bring all the benefits of probiotics, but without the practical problem of adding live bacteria to the diet. Furthermore, probiotics do not colonise the gut and so are washed out usually after a few days.

Prebiotics are mostly sugar-like compounds with a degree of polymerization of between two and 60. They comprise mainly fructose and galactose polymers and occur naturally in the diet in artichokes, the leek, onion and garlic family, in legumes and in small amounts in cereals. They are also a significant component of breast milk. They largely escape digestion in the small bowel, are fermented and produce gas and short chain fatty acids. Their laxative properties, unlike non-starch polysaccharides and resistant starch, are minimal. Oligofructose

TABLE 2 Claimed health benefits of prebiotics

Fermented and provide short chain fatty acids and biomass, but not significant laxation
Selectively affect the balance of the flora — may enhance resistance to gut infection
Simulate Vitamin B synthesis
Increase calcium absorption
Anti cancer effects (in animal models)

appears to increase calcium absorption (Table 2). Prebiotics are currently attracting a lot of interest and are the subject of several reviews (Hidaka et al 1986, Schrezenmeir & de Vrese 2001, Roberfroid & Gibson 2002). They are simple molecules that can readily be made using industrial enzymic processes from cheap raw materials, and are likely to be the designer foods of the future. They come into the category of functional foods or nutraceuticals.

The key, essential and distinguishing property of prebiotics is their capacity to selectively stimulate the growth of species of bacteria in the genera bifidobacterium and lactobacillus. The original demonstration of their effect was probably by Japanese scientists, reported at the First Conference of the Japan Bifidus Foundation in May 1981 and written up in the first issue of a then new journal, *Bifidobacteria and Microflora* in 1982. A summary of some of the subsequent human studies in given in Table 3.

TABLE 3 Studies in humans of the prebiotic effect of carbohydrates

				Bifodobacteria/ bacteroides ratio in faeces		
Prebiotic	*Amount (g/d)*	*Duration of feeding (d)*	*Number of subjects*	*Control*	*Test*	*Source*
Oligofructose	8	14	23-16	0.10	0.40	Hidaka et al (1986)
Galacto-oligo saccharides	2.5	7	12	1.07	1.58	Ito et al (1990)
	5.0	7	12	1.07	1.35	
	10.0	7	12	1.07	2.75	
Oligofructose	15	15	8	0.25	5.01	Gibson et al (1995)
Inulin	15	15	4	0.32	1.99	Kleessen et al (1997)
	20	9	10	0.08	0.40	
	40	9	10	0.08	3.98	
Oligofructose	5	21	8	0.13	0.42	Rao (2001)
Inulin	8	14	10	0.25	0.63	Tuohy et al (2001)

These studies are ones where it has been possible to calculate a ratio of bifidobacteria to bacteroides in faecal flora during a control (or initial) period and then with a prebiotic. There are many other studies where prebiotics have been fed and either microbial or other effects measured (see reviews cited above). The papers shown in Table 3 illustrate the wide variation in the control or initial ratio of bifidobacteria to bacteroides from 0.08 to 1.07, but mostly less than 1.0, in the planktonic flora, and the equally wide range in responses with final ratios from 0.4 to 5.01, representing changes of up to, but rarely more than 1 $\log_{10}$ in bifidobacterial counts. Average values for the data in Table 3 are control 0.44 and prebiotic 1.84, indicating that bifidobacteria move from being 44% of the proportion of bacteroides to 1.84 times as many, and become the dominant flora in faeces. This shift in the balance of the flora would appear to be real, and recaptures to some extent the pattern of flora of breast-fed infants.

Selectivity

Any carbohydrate that reaches the large bowel will provide a substrate for microbial growth and metabolism. The early model for this effect was dietary fibre, or the non-starch-polysaccharides (NSPs). Studies of the effect of NSPs on the flora have shown little that was remarkable apart from some general increase in biomass. Prebiotics must, however, selectively increase bifidobacteria or lactobacilli. To demonstrate this it is therefore essential to measure the other major genera in faeces, or wherever the effect is being looked for, and show selectivity for growth of the probiotic-like bacteria. Selectivity does not rule out an increase in other genera, just a greater increase in the probiotic species. For example, in a recent paper by Tuohy et al (2001) bifidobacteria increased by only $\log_{10}$ 0.2 cells/g faeces after 14 days on inulin, but bacteroides went down $\log_{10}$ 0.2 cells/g faeces giving a change in the proportion of bifidobacteria to bacteroides from 25% to 63%. The mechanism is relatively easy to understand for bifidobacteria since they have relatively high amounts of β-fructosidase that is selective for the β1–2 glycosidic bond present in oligosaccharides, but more difficult to understand for lactobacilli and the effect of prebiotics such as galactooligosaccharides.

Mucosa-associated flora

Whilst the interpretation of the role of commensal flora in the gut has for long been based on characterizing the planktonic or luminal bacteria, it is now evident that a mucosa-associated flora (MAF) exists and is more likely to interact directly with, and influence, the epithelial cells and mucosa-associated lymphoid tissue. There are currently only two studies that address the issue of prebiotics and MAF, one in animals, and the other in humans undergoing colonoscopy.

In Kleessen et al's (2003) study rats were fed a mixture of oligofructose and long chain inulin or a standard chow diet. The prebiotic diet resulted in higher villi and deeper crypts in bacteria-associated, but not germ-free rats and a thicker mucus layer. Mucosa-associated bifidobacteria increased with the prebiotic whereas bacteroides did not change.

In a study by Langlands et al (2004) subjects attending for routine colonoscopy were randomized to receive either 7.5 g oligofructose and 7.5 g inulin, or placebo for 14 days prior to attendance for endoscopy. Multiple biopsies were taken and showed a difference in the ratio of bifidobacteria to bacteroides on the mucosa of 1.6×10^{-3} (placebo) versus 1.0×10^{-2} (probiotic) in the proximal colon and 7.9×10^{-4} (placebo) and 7.9×10^{-3} (prebiotic) in the distal colon. There were also significant increases in lactobacilli and eubacteria. Although bifidobacteria do not come to be the dominant flora on the mucosa in this study, unlike studies of the planktonic populations, nevertheless it is reassuring to know that this microenvironment, which is so important to mucosal and immune function, can be affected by a small dietary change.

With new molecular techniques emerging for gut microbiology, much better profiling of the population is possible. This is almost certain to turn up anomalies in the selectivity concept as previously unconsidered genera show changes with prebiotics (Langlands et al 2004).

Clinical benefit of prebiotics

At the present time there is virtually no literature on this subject. Pro- and prebiotic combinations have been used, but it is not possible to determine from such studies a unique role for prebiotics.

Welters et al (2002) have, however, used inulin to treat pouchitis, a condition that responds well to probiotics. In a randomized double-blind controlled study they gave 20 patients a large dose (24 g/day) of inulin for three weeks and observed a lowering of pH from 5.62 to 5.33, a reduction in deoxycholic acid concentrations and some reduction in *B. fragilis* numbers, but no other significant changes in the flora. Importantly, however, the pouchitis disease activity index fell significantly from 5.39 (placebo) to 4.05 (inulin) and they conclude that inulin is beneficial.

In a clinical study of travellers' diarrhoea (Cummings et al 2001) 10 g/day of oligofructose was used in a randomised control-led trial of 244 subjects to prevent diarrhoea whilst on holiday. Although there was no difference in overall stool frequency between the placebo and oligofructose groups, the severity of attacks was reduced by oligofructose (Fig. 1).

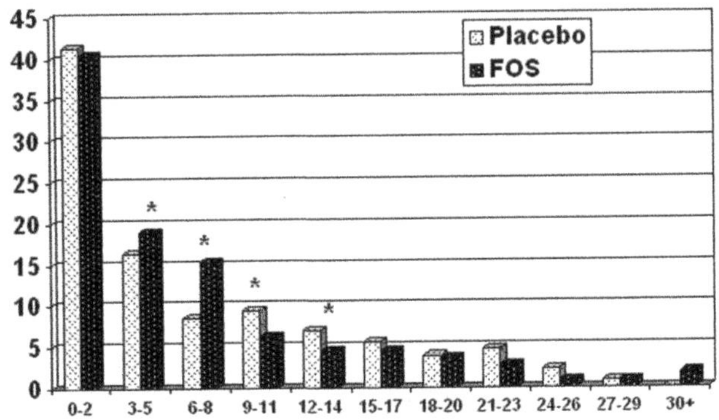

FIG. 1. Distribution of diarrhoea severity score in travellers ($n = 244$) taking either a placebo or oligofructose 10 g/d for 14 days prior to travel and during the holiday. The figure shows that the severity of episodes of diarrhoea was less with the oligofructose shifting the distribution of scores to the left (Cummings et al 2001). $P < 0.05$.

Antibiotics

Ulcerative colitis

In recent years, more than a dozen randomized controlled trials (RCTs) of antibiotic use in ulcerative colitis (UC) have been published. Six antibiotics have been used either singly or in combination. Trials which showed positive results either initially or after a prolonged period of treatment with antibiotics are summarized in Table 4.

The first RCT of antibiotics to treat UC was done by Dickinson et al (1985) using vancomycin (2 g given orally per day), which was chosen because of its benefits in antibiotic-associated diarrhoea. In 33 cases of acute UC, there was no clear benefit although fewer came to surgery in the antibiotic group (2/18) compared to the placebo group (7/15).

Gilat et al (1987, 1989) treated 42 UC patients in an acute attack for 28 days with 1.35 g metronidazole per day by mouth, and found no benefit over treatment with sulfasalazine in an RCT. Only 26% improved with antibiotic, versus 68% with sulfasalazine. They then followed the patients for one year, giving 0.6 g metronidazole, or 2 g sulfasalazine per day. Metronidazole was found to be 'slightly more effective' after 12 months of treatment.

Tobramycin appears to be successful when used in acute attacks. Burke et al (1990) found 74% of patients in remission at 28 days, following a 7 day course of oral antibiotic, compared with 43% on placebo. However, the same group found no benefit for tobramycin at 12 and 24 months in 81 patients (Lobo et al 1993). Overall, therefore, tobramycin showed no consistent benefit.

TABLE 4 Clinical trials of antibiotic use in ulcerative colitis

Antibiotic	*Dose/day*	*Route*	*Duration*	*n*	*Study design*	*Results*	*Source*
Metronidazole	1.35 g	Oral	28 days	42	RCT v. salazopyrine 4.5 g. Acute attack	No benefit	Gilat et al (1987)
Metronidazole	0.6 g	Oral	12 months	33	RCT v. salazopyrine 2 g. Maintenance of remission	Benefit at 12 months, also longer remission	Gilat et al (1989)
Tobramycin		Oral	7 days	84	RCT Acute UC	At 28 days tobramycin better	Burke et al (1990)
Tobramycin		Oral	7 days	81	RCT Long-term follow up	No benefit at 12 or 24 months	Lobo et al (1993)
Vancomycin	2 g	Oral	7 days	33	RCT Acute UC	No clear benefit but less surgery	Dickinson et al (1985)
Ciprofloxacin	1–1.5 g	Oral	6 months	83	RCT	Benefit at 6 months, but not at 12	Turunen et al (1998)
Amoxicillin-clavulanic acid	2.25 g	Oral	5 days	30	RCT Acute attack	Release of inflammatory mediators reduced	Casellas et al (1998)
Rifaximin	800 mg	Oral	10 days	28	RCT Moderate to severe steroid resistant UC	Substantial clinical and endoscopic benefit	Gionchetti et al (1999b)
Rifaximin	800 mg	Oral	10 days	31	Open label Active UC	Effective in patients with mild to moderate left-sided UC	Lukas et al (2002)

RCT, randomized controlled trial; UC, ulcerative colitis.

Whilst the objective of using antibiotics in UC is to change the microflora beneficially, it is surprising that very few studies include reports of any bacteriology done on the patients. Turunen et al (1998) performed semi-quantitative estimates of mucosal bacteria from endoscopic biopsies. They reported no enteric pathogens at the start of the study, and the disappearance of Gram-negative facultative anaerobes in the antibiotic (ciprofloxacin) group. After six months of ciprofloxacin treatment, the failure rate was 21% in the treated group, and 44% with the placebo ($P=0.02$). However, endoscopic and histological changes were no different between these groups at six months, and there was no clinical benefit at one year.

Casellas et al (1998) gave amoxicillin/clavulanic acid orally for 5 days to UC patients who had an acute attack without apparent benefit, except in the release of inflammatory mediators, quantitated by mucosal release of eicosanoids using rectal dialysis.

Rifaximin is a new generation antibiotic targeted at the gut, that is derived from rifamycin. It is non-absorbable resulting in high concentrations in stool and has a broad spectrum of activity against both Gram-positive and Gram-negative bacteria, as well as colonic anaerobes. This drug has been used successfully to treat pouchitis (Gionchetti et al 1999a) and there have been two trials in UC patients (Gionchetti et al 1999b, Lukas et al 2002). In an open label study, 31 patients with mild to moderate, predominately left-sided UC, took 400 mg twice daily for 10 days. At 28 days, the clinical activity index and sigmoidoscopy scores were significantly better, and only two patients were worse (Lukas et al 2002). In the Gionchetti et al (1999b) RCT of 28 patients, the same dose was used in moderate to severely affected, steroid-resistant UC patients. Sixty-four percent of treated patients were substantially better as defined by an improvement of the disease's clinical activity compared to 42% of the placebo group, and significant improvement was seen in stool frequency, rectal bleeding and sigmoidoscopy scores. At present, rifaximin has the best record for treating acute UC but it should be noted that the studies have involved small numbers of patients only. Thus long-term studies with larger patient groups are clearly needed to fully evaluate and verify its role in the treatment of UC.

Crohn's disease

The antibiotics used in the treatment of Crohn's disease can be divided into two groups: broad-spectrum antibiotics and antimycobacterial therapy.

Broad-spectrum antibiotics

Metronidazole works best when used to treat perianal Crohn's disease and in the prevention of Crohn's disease recurrence post surgery. In a study by Bernstein et al

(1980) in which patients with chronic unremitting Crohn's disease were treated with metronidazole, 10 of 18 patients had complete healing of their disease. In a follow-up study, all patients who had recurrence of their symptoms when the metronidazole was discontinued, promptly responded to reinstitution of full-dose therapy (Brandt et al 1982).

Ornidazole, a nitroimidazole like metronidazole, was given at 1 g per day in a study by Rutgeerts et al (2002) to prevent post operative recurrence of Crohn's disease. Eighty patients received either ornidazole or placebo for 1 year. A significant difference was observed between the two groups at 1 year. Endoscopic recurrence was present in 54% of patients in the antibiotic group versus 79% of the patients who received the placebo, and clinical recurrence was 8% verus 37%, respectively. However, the difference was not statistically significant after 2 and 3 years. Ornidazole was poorly tolerated and 8 of 40 patients developed neuropathy.

Ciprofloxacin has been used singly or in combination with metronidazole. In a small study by Peppercorn (Peppercorn 1993), all the patients had a dramatic improvement in abdominal pain and diarrhoea coincident with institution of the ciprofloxacin. Greenbloom et al (1998) treated 72 patients with a combination of ciprofloxacin and metronidazole. Clinical remission was observed in 49 patients (68%) and 55 patients (76%) showed a clinical response. Subgroup analysis however showed that a clinical response was noted in 29 of 43 patients who were not on concurrent steroids compared to 26 of 29 patients on steroids.

Antimycobacterial therapy

Antimycobacterial therapy has been used in the treatment of Crohn's disease because mycobacteria have been isolated from Crohn's tissues. Histologically Crohn's tissues resemble tissues from animals with Johne's disease which is caused by *Mycobacterium avium paratuberculosis*.

Few clinical trials using antimycobacterial therapy in Crohn's disease have shown clear benefit. Prantera et al (1994) achieved good results with quadruple therapy consisting of ethambutol, clofazimine, dapsone and rifampicin. The study group was patients with refractory, steroid-dependent Crohn's disease. Three patients out of 19 on the active treatment relapsed compared with 11 of 17 patients on placebo. Nine patients whose disease relapsed or persisted on placebo were crossed over to active drug treatment. Five of these patients achieved sustained remission. However, substantial endoscopic or radiological healing did not occur.

Meta-analysis of 6 trials by Borgaonkar et al (2000) showed that the pooled odds ratio for maintenance of remission in treatment versus control was 1.1 (not significant). Subgroup analysis of two trials which used antimycobacterial therapy in combination with corticosteroids to induce remission, followed by maintenance therapy with antimycobacterials, yielded a pooled odds ratio of maintenance of remission in treatment versus controls of 3.37 (95% confidence interval [CI] 1.38–8.24) in favour of treatment, which was statistically significant. Subgroup analysis of four trials which did not use corticosteroids to induce remission yielded a pooled odds ratio of maintenance of remission in treatment versus controls of 0.69 (95% CI 0.39–1.21), which was not significant.

In a study by Gui et al (1997) 46 patients were treated with rifabutin in combination with a macrolide antibiotic (clarithromycin or azithromycin). Patients were treated for a mean of 18.7 (range 6–35) months and followed up for 25.1 (range 7–41) months. Of the 19 patients who were steroid dependent at the start of the study, only two continued to require steroids when treatment was established. A reduction in the Harvey-Bradshaw Crohn's disease activity index occurred after 6 month's treatment ($P=0.004$, paired Wilcoxon test) and was maintained at 24 months ($P<0.001$). An improvement in inflammatory parameters was observed as measured by a reduction in erythrocyte sedimentation rate ($P=0.009$) and C-reactive protein ($P=0.03$) at 18 months compared with pre-treatment levels, and an increase in serum albumin at 12 months ($P=0.04$). When subsets of the study population were analysed, patients with pan-intestinal disease achieved better remission at two years than those with less extensive involvement ($P=0.04$, Mann–Whitney U-test). These data need to be confirmed with a large randomized double-blind control trial.

There are various reasons why not all patients respond to antibiotics and the use of antibiotics is not more widespread. Some of the reasons are listed below:

- Bacteria involved not known, therefore no antibiotic sensitivities.
- Bacteria may be part of a biofilm consortium on the mucosal surface, making the bacteria more resistant to antibiotic treatment.
- Therapeutic levels of antibiotics may not be reached at target site.
- Resistance developing, especially with long term use.
- Systemic side-effects.

Thus more work will have to be done to explore the above issues further before antibiotics will be considered as part of the routine armoury of treatment for inflammatory bowel disease.

Conclusion

Probiotics, prebiotics and antibiotics are all potential treatments for IBD that exploit the key role that bacteria play in either the initiation or perpetuation of Crohn's and UC. Pro and prebiotics offer major advantages to patients in terms of safety and absence of toxicity, and they are likely to be cheap in the future.

However, good RCTs are needed of probiotics supported by laboratory studies to identify mechanisms and thus guide the choice of probiotics that will work best.

Prebiotics are an exciting new development that may find a place in enhancing mucosal defences and may also enhance probiotic therapy in IBD. Their capacity to alter the balance of the MAF means that they could be key to controlling the immune response to commensal flora.

The full potential of antibiotics in IBD has not yet been exploited for a number of reasons. Not least is that we do not know which bacteria to target. Furthermore, these bacteria live in biofilms where antibiotic resistance is common, and require novel approaches as was discovered in the treatment of *Helicobacter pylori*. Disturbance to the overall balance of the flora may limit antibiotic use in the long term.

References

Bernstein LH, Frank MS, Brandt LJ, Boley SJ 1980 Healing of perineal Crohn's disease with metronidazole. Gastroenterology 79:357–365

Borgaonkar MR, MacIntosh DG, Fardy JM 2000 A meta-analysis of antimycobacterial therapy for Crohn's disease. Am J Gastroenterol 95:725–729

Brandt LJ, Bernstein LH, Boley SJ, Frank MS 1982 Metronidazole therapy for perineal Crohn's disease: a follow up study. Gastroenterology 83:383–387

Burke DA, Axon ATR, Clayden SA et al 1990 The efficacy of tobramycin in the treatment of ulcerative colitis. Aliment Pharmacol Ther 4:123–129

Casellas F, Borruel N, Papo M et al 1998 Antiinflammatory effects of enterically coated amoxicillin-clavulanic acid in active ulcerative colitis. Inflamm Bowel Dis 4:1–5

Cremonini F, Di Caro S, Nista EC et al 2002 Meta-analysis: the effect of probiotic administration on antibiotic-associated diarrhoea. Aliment Pharmacol Ther 16:1461–1467

Cummings JH, Christie S, Cole TJ 2001 A study of fructo oligosaccharides in the prevention of travellers' diarrhoea. Aliment Pharmacol Ther 15:1139–1145

de Roos NM, Katan MB 2000 Effects of probiotic bacteria on diarrhea, lipid metabolism, and carcinogenesis: a review of papers published between 1988 and 1998. Am J Clin Nutr 71:405–411

Dickinson RJ, O'Connor HJ, Pinder I 1985 Double-blind controlled trial of oral vancomycin as adjunctive treatment in acute exacerbations of idiopathic colitis. Gut 26:1380–1384

D'Souza AL, Rajkumar C, Cooke J, Bulpitt CJ 2002 Probiotics in prevention of antibiotic associated diarrhoea: meta-analysis. Br Med J 324:1361–1364

Gibson GR, Beatty ER, Wang X, Cummings JH 1995 Selective stimulation of Bifidobacteria in the human colon by oligofructose and inulin. Gastroenterology 108:975–982

Gilat T, Suissa A, Leichtman G et al 1987 A comparative study of metronidazole and sulfasalazine in active, not severe, ulcerative colitis. An Israeli multicenter trial. J Clin Gastroenterol 9:415–417

Gilat T, Leichtman G, Delpre G et al 1989 A comparison of metronidazole and sulfasalazine in the maintenance of remission in patients with ulcerative colitis. J Clin Gastroenterol 11:392–395

Gionchetti P, Rizzello F, Venturi A et al 1999a Antibiotic combination therapy in patients with chronic, treatment-resistant pouchitis. Aliment Pharmacol Ther 13:713–718

Gionchetti P, Rizzello F, Venturi A et al 1999b Rifaxamin in patients with moderate or severe ulcerative colitis refractory to steroid treatment: a double-blind, placebo-controlled trial. Dig Dis Sci 44:1220–1221
Greenbloom SL, Steinhart AH, Greenberg GR 1998 Combination ciprofloxacin and metronidazole for active Crohn's disease. Can J Gastroenterol 12:53–56
Gui GP, Thomas PR, Tizard ML et al 1997 Two-year-outcomes analysis of Crohn's disease treated with rifabutin and macrolide antibiotics. J Antimicrob Chemother 39:393–400
Hart AL, Kamm MA 2003 Use of probiotics in the treatment of inflammatory bowel disease. J Clin Gastroenterol 36:111–119
Hidaka H, Eida T, Takizawa T, Tokunaga T, Tashiro Y 1986 Effects of fructooligosaccharides on intestinal flora and human health. Bifidobacteria Microflora 5:37–50
Ito M, Deguchi Y, Miyamori A et al 1990 Effects of administration of galactooligosaccharides on the human faecal microflora, stool weight and abdominal sensation. Microb Ecol Health Dis 3:285–292
Jonkers D, Stockbrugger R 2003 Probiotics and inflammatory bowel disease. J R Soc Med 96:167–171
Kalliomaki M, Salminen S, Poussa T, Arvilommi H, Isolauri E 2003 Probiotics and prevention of atopic disease: 4-year follow-up of a randomised placebo-controlled trial. Lancet 361:1869–1871
Kleessen B, Sykura B, Zunft H-J, Blaut M 1997 Effects of inulin and lactose on fecal microflora, microbial activity, and bowel habit in elderly constipated persons. Am J Clin Nutr 65:1397–1492
Kleessen B, Hartmann L, Blaut M 2003 Fructans in the diet cause alterations of intestinal mucosal architecture, released mucins and mucosa-associated bifidobacteria in gnotobiotic rats. Br J Nutr 89:597–606
Langlands SJ, Hopkins MJ, Coleman N, Cummings JH 2004 Prebiotic carbohydrates modify the mucosa-associated microflora of the human large bowel. Gut, in press
Lobo AJ, Burke D, Sobala GM, Axon ATR 1993 Oral tobramycin in ulcerative colitis: effect on maintenance of remission. Aliment Pharmacol Ther 7:155–158
Lukas M, Konecny M, Zboril V 2002 Rifaximin in patients with mild to moderate activity of ulcerative colitis: An open label study. Gastroenterology 122:A434
Marteau PR 2002 Probiotics in clinical conditions. Clin Rev Allergy Immunol 22:255–273
Peppercorn MA 1993 Is there a role for antibiotics as primary therapy in Crohn's ileitis? Gastroenterology 17:235–237
Prantera C, Kohn A, Mangiarotti R, Andreoli A, Luzi C 1994 Antimycobacterial therapy in Crohn's disease: results of a controlled, double-blind trial with a multiple antibiotic regimen. Am J Gastroenterol 89:513–518
Rao VA 2001 The prebiotic properties of oligofructose at low intake levels. Nutr Res 21:843–848
Reid G, Jass J, Sebulsky MT, McCormick JK 2003 Potential uses of probiotics in clinical practice. Clin Microbiol Rev 16:658–672
Roberfroid M, Gibson G 2002 Nutritional and health benefits of inulin and oligofructose. Br J Nutr 87:S139–S311
Rutgeerts P, van Assche G, D'Haens G et al 2002 Ornidazole for prophylaxis of post operative recurrence of Crohn's disease: final results of a double blind placebo controlled trial. Gastroenterology 122:666
Schrezenmeir J, de Vrese M 2001 Probiotics, prebiotics, and synbiotics—approaching a definition. Am J Clin Nutr 73:361S–364S
Stephen A, Henry J, Marks J, Shortt C 2002 Probiotics and health. Br J Nutr 87:3–99
Szajewska H, Mrukowicz JZ 2001 Probiotics in the treatment and prevention of acute infectious diarrhea in infants and children: a systematic review of published randomized, double-blind, placebo-controlled trials. J Pediatr Gastroenterol Nutr 33(suppl 2):S17–S25

Tuohy KM, Finley RK, Wynne AG, Gibson GR 2001 A human volunteer study on the prebiotic effects of HP-Inulin—faecal bacteria enumerated using fluorescent *in situ* hybridisation (FISH). Anaerobe 7:113–118

Turunen UM, Farkkila MA, Hakala K et al 1998 Long-term treatment of ulcerative colitis with ciprofloxacin: a prospective, double-blind, placebo-controlled study. Gastroenterology 115:1072–1078

Van Niel CW, Feudtner C, Garrison MM, Christakis DA 2002 *Lactobacillus* therapy for acute infectious diarrhea in children: a meta-analysis. Pediatrics 109:678–684

Welters CFM, Heineman E, Thunnissen BJM et al 2002 Effect of dietary inulin supplementation on inflammation of pouch mucosa in patients with an ileal pouch-anal anastomosis. Dis Colon Rectum 45:621–627

DISCUSSION

Gibson: In your prebiotic treatments, you are giving fructo-oligosaccharides (FOS), yet we have got lots of these in our diets already. Wheat is the major source. The problem is that what you may be doing is giving just a drop in the ocean. Mainly, you are giving a fermentation product. Bifidobacteria don't take up probiotics and produce gas: they metabolise them in a different way. The problem with using FOS as a prebiotic is that you don't know what the baseline diet is and you will get heterogeneity in response. I feel that the future is not in FOS for prebiotics, but more in the designer prebiotics that we know will not be in the diet. Lactulose is a very good prebiotic, but you only need 5 g a day. If you have 10 g you may experience fermentation effects, such as diarrhoea and bloating.

Cummings: There are a lot of issues here; I'll touch on a couple. I agree that designer carbohydrates will be a big story, and FOS may not be the best candidate. FOS is present in the diet, but our dietary studies suggest that there is not that much of it. In the UK it is around 5–7 g per day. We do have to take account of this if we are doing studies. If you are doing a feeding study you need to produce a diet minimizing FOS content. This is what we did with our original study. You are right in saying that perhaps this is why people have failed with these sorts of studies: because they haven't controlled intake of foods such as onions and cereals. The second point is that whilst prebiotics are largely fermented in the right side of the colon, why do they affect events more distally? This may be why they don't have an effect on bowel habit. But we can design them to be more slowly fermented. There are forms of inulin that have longer chain lengths that are now on the market. These are more slowly fermented and seem to have more of an effect further round the bowel. Finally, I agree that for all prebiotics, intake will be limited potentially by gas production.

Jewell: If you use synbiotics, when you stop the probiotic bit but continue with the prebiotic, do the probiotic organisms in the stool decline?

Cummings: Unfortunately we don't know the answer to that.

Furrie: This should promote the probiotic organisms in theory. Also, you will promote the already existing bifidobacteria. One idea that is promising is to give a hit of probiotic and then maintain people on a prebiotic.

Kamm: The Bologna group (Campieri and colleagues) have looked at the persistence of probiotic bacteria in a clinical trial situation. In most patients they drop off very quickly and they are not identifiable. However, in a few subjects you can still find the presence of these identifiable organisms two or three months later.

Jewell: Presumably this could be a question of the background diet, or some kind of genetic host factor.

Sartor: I think the mechanisms underlying the positive effects of probiotics are very much up in the air. There are three main possibilities: colonization resistance, improvement in epithelial health through the short-chain fatty acid production, and the altered cytokine ratios. In cell cultures and in some *in vivo* studies it is possible to decrease tumour necrosis factor (TNF) and increase interleukin (IL)10 production. The interesting thing is that you don't need viable bacteria in order to alter the cytokine ratio. Either protein extracts or DNA will also do this. I like the concept of prebiotics because this is probably the most physiological and cheapest approach. Regarding the studies you showed, they involved a high sucrose pre-existing diet that increased bacteroides and then they changed to oligosaccharides. What happens if you run a low sucrose baseline diet and then give the oligosaccharides? Is it just the difference between the short chain saccharides versus the oligosaccharides? Have you had the opportunity to look at short chain fatty acid production, which may be very important?

Cummings: Yes, we have looked at short chain fatty acids both in faeces and in our *in vitro* models. Prebiotics are a source of these, but they are not an especially good source of butyrate as, for example, is starch. This has held up in virtually all the studies that have looked at this, so we have lost interest in this mechanism. As far as sucrose is concerned, sucrose intakes in the UK and Europe are in the range of 80–100 g/day for men and 60–80 g/day for women. We changed only 15 g in the early study I showed, so I don't think this is a significant change in sucrose intake. Are there any low sucrose populations that get IBD? I don't know of any. Most unaccultured populations that have low sucrose intakes also have very low IBD risk.

Sartor: Not only will there be a parasite difference, but also there will be extreme dietary differences between third world and western populations. Is it the presence of sucrose that is significant, or other factors that are altering the bacterial milieu? You are focusing on the oligosaccharides, but can one consider fibre as a prebiotic? What about the polysaccharides — the non absorbable food stuffs?

Cummings: People are starting to claim that some of the non-starch polysaccharides are prebiotic, but the studies that have been published so far have not been clean. There have been mixtures of carbohydrates fed and the range of bacteria measured has been insufficient to demonstrate a selective effect. Some publications just look at bifidobacteria and lactobacilli. My understanding is that bifidobacteria do well on fructans because they express high amounts of β-fructosidase that gives them a selective advantage. However, I don't understand why they do well on galacto-oligosaccharides. Is it the same enzyme that is involved? I don't know enough microbial biochemistry to predict whether some of the bigger fibre molecules will be prebiotic. I would be surprised if they were. The studies we have done in the past indicate that they cause an across the board increase in bacteria rather than a selective one.

Sartor: The group associated with the Kirin brewing company in Tokyo have reported a small trial with prebiotics in UC. Their prebiotic is from the residue from their brewing process! They claim excellent effects in several animal models and UC patients.

Cummings: If you look at what they are feeding it is a mixture of molecular weights. It may be an interesting study from their point of view but it doesn't help us with prebiotics.

Furrie: We have been doing feeding trials on elderly subjects in which we are using synbiotics to improve quality of life. In the elderly, constipation and slowing of transit time is a serious problem. We find that in the feeding studies the patients respond far better to fructo-oligosaccharides than galacto-oligosaccharides.

Fedorak: Using such methodologies have you been able to quantitate the magnitude of changes using prebiotics and probiotics? If you are giving both a prebiotic and probiotic, will you get an increased, or the same, magnitude of change? Is a prebiotic just a weak probiotic?

Cummings: It is probably the other way round. With probiotics it is usual to give a single species and the counts can get up to 10^8 or 10^9 in faeces. This is beginning to comprise up to 1% of the flora. But with prebiotics, bifididobacteria can reach 70–80% of the flora.

Fedorak: Have you measured the small bowel effects of the prebiotics? Do you see changes proximally?

Cummings: We haven't done this, and there is no a priori reason why there should be an effect other than an osmotic one. I wouldn't rule it out though. There is one paper in the literature using stable isotopes showing that FOS improves Ca^{2+} absorption (van den Heuvel et al 1999).

Meddings: I'm puzzled: how can this increase Ca^{2+} absorption?

Cummings: I have no idea. I have been at meetings where this has been debated, but it isn't clear. Some people have suggested it is a colonic effect, although my understanding is that the human colon does not express Ca^{2+} binding protein. However, if this is true, it has huge implications, particularly in adolescence.

Uhlig: In the early colonization phase in newborn babies there are high numbers of bifidobacteria. Are there endogenous sources which initiate and mimic this prebiotic effect?

Cummings: There are naturally occurring oligosaccharides in the diet. The cow's milk we all drink is not a good source, but in breast milk there is a hugely dominating and fascinating range of oligoscaccharides, resulting in the flora of a baby being dominated by bifididobacteria. Oligosaccharides are present in significant quantities in artichokes, and in moderate quantities in onion, garlic and leeks. There are small amounts in cereals.

Ghosh: Going back to the issue of mechanism of action, do we know whether the new designer carbohydrates, such as manno-oligos, have primary immunomodulatory roles? These might bind to the mannose receptor family on dendritic cells. When we ingest the oligos do they form glyco-conjugates in the gut?

Cummings: We have studies ongoing demonstrating that there is an interaction between oligosaccharides and the cell surface of gastrointestinal-type cells. This would then manifest itself as resistance to invasion and cytotoxicity of pathogens. There is an interaction. In the animal world they think it is a very important interaction.

Reference

van den Heuvel EGHM, Muys T, van Dokkum W, Schaafsma G 1999 Oligofructose stimulates calcium absorption in adolescents. Am J Clin Nutrit 69:544–548

The epithelium in inflammatory bowel disease: potential role of endocytosis of junctional proteins in barrier disruption

Andrei I. Ivanov, Asma Nusrat and Charles A. Parkos[1]

Epithelial Pathobiology Research Unit, Department of Pathology and Laboratory Medicine, Whitehead Research Building, Emory University, 615 Michael Street, Atlanta, GA 30322, USA

Abstract. A key feature of inflammatory bowel disease (IBD) is disruption of the intestinal epithelial barrier by unknown mechanisms. Integrity of the epithelial barrier is determined by an apical junctional complex that is composed of tight junction (TJ) and adherens junction (AJ). Previous observations have suggested that alterations in the apical junctional complex occur in IBD. Localization studies in mucosal biopsies from IBD patients have revealed disappearance of key TJ (occludin, JAM1, ZO1, claudin 1) and AJ (E-cadherin, β-catenin) proteins from intercellular junctions. *In vitro* experiments examining the effects of inflammatory cytokines on model intestinal epithelial monolayers suggest that disruption of the epithelial barrier is associated with internalization of transmembrane TJ proteins, JAM1, occludin and claudins 1/4. The mechanism(s) of internalization of intercellular junctions can be modelled *in vitro* by calcium depletion of confluent epithelial cell monolayers. Using this model, we have observed rapid, orchestrated endocytosis of all AJ and TJ proteins into a subapical cytoplasmic compartment that is independent of caveolae/lipid rafts and macropinocytosis. However, inhibitors of clathrin-mediated endocytosis effectively block internalization of AJs and TJs, and junctional proteins colocalize with clathrin. Interestingly, internalized AJ and TJ proteins enter early endosomes followed by movement to organelles that do not label with markers of late and recycling endosomes, lysosomes or Golgi but appear to represent a unique storage compartment that colocalizes with t-SNARE protein, syntaxin 4. A better understanding of the mechanisms of junctional internalization and recycling will likely provide new insights into the mechanisms of altered barrier function in IBD.

2004 Inflammatory bowel disease — crossroads of microbes, epithelium and immune systems. Wiley, Chichester (Novartis Foundation Symposium 263) p 115–132

[1]This paper was presented at the symposium by Charles A. Parkos to whom correspondence should be addressed.

Epithelial permeability and intercellular junctions are altered in IBD

Inflammatory bowel disease (IBD), consisting of ulcerative colitis (UC) and Crohn's disease (CD), represents a clinical syndrome characterized, in part, by relapsing diarrhoea and severe mucosal inflammation. While the underlying mechanisms behind the pathogenesis of IBD are not understood, it is well documented in patients with CD and UC (Meddings 1997, Tibble & Bjarnason 2001) as well as in animal models of IBD (Hermiston & Gordon 1995, Panwala et al 1998) that enhanced epithelial paracellular permeability may play a role. Before summarizing the evidence, it is worth briefly reviewing structural components involved in regulation of intestinal epithelial permeability.

Paracellular permeability across a polarized epithelial cell monolayer, such as that lining the intestine, is regulated by the apical-most intercellular junction referred to as the apical junctional complex (AJC). The major constituents of the AJC are the tight junction (TJ) and the subjacent adherens junction (AJ). The TJ acts as a gate regulating solute flux and a fence preventing lateral diffusion of molecules between the apical and basolateral plasma membrane domains (Tsukita et al 2001). Freeze-fracture replica electron microscopic analyses of epithelial cells reveal the TJ as a meshwork of anastomosing intramembranous strands of particles in which strand complexity correlates with leakiness of the barrier. While the TJ is responsible for maintaining the seal between epithelial cells (Madara 1998), the AJ is vital for initiating and maintaining cell–cell contacts (Yap et al 1997). Both of these junctions represent multiprotein complexes composed of transmembrane proteins and cytosolic plaque proteins (Yap et al 1997, Tsukita et al 2001). The former proteins mediate cell–cell adhesion whereas the latter link the TJ and AJ to the cytoskeleton and participate in intracellular signalling. Transmembrane protein components of the TJ include occludin, claudins and junctional adhesion molecules (JAMs) while cytoplasmic plaque proteins consist of a number of scaffolding and signalling molecules such as the zonula occludens (ZO) protein family (Tsukita et al 2001). One of the major transmembrane protein constituents of the AJ in epithelial cells is E-cadherin. The extracellular domain of this protein binds multiple Ca^{2+} cations resulting in a unique, adhesion-promoting conformational change. The cytoplasmic face of the AJ is comprised of several members of the catenin protein family that interact with E-cadherin (Yap et al 1997).

Because integrity of the TJ and AJ determines normal barrier function in epithelia, it has been hypothesized that defects of AJC structure may underly increased mucosal permeability observed in patients with IBD (Hermiston & Gordon 1995). Initial experimental support for this hypothesis came from observations on the distribution of E-cadherin in human IBD tissues (Jankowski et al 1998). Expression of E-cadherin was down-regulated in colonic epithelial cells of UC and CD patients. Morphological analyses revealed that E-cadherin was focally absent from intercellular junctions and was detected in the cytoplasm. In a subsequent investigation, it was reported that TJ structure was altered in IBD (Schmitz et al 1999). Freeze-fracture electron microscopic analyses clearly documented a decrease in the number and complexity of TJ strands in colonic biopsies from UC patients (Schmitz et al 1999). Two recent papers have provided additional insights into mechanisms underlying alteration of AJC structure in IBD. In particular, immunohistochemical analysis (Kucharzik et al 2001) revealed substantial loss of occludin from TJs in colonic mucosa of patients with UC and CD. Interestingly, the loss of occludin was observed in crypt epithelial cells away from sites of active transmigration of neutrophils and crypt abscess formation. The loss of occludin from intercellular junctions correlated with down-regulation of expression at the mRNA and protein level. It was also observed that IBD biopsies had diminished levels of other TJ (ZO1, JAM1, claudin 1) and AJ (E-cadherin, β-catenin) proteins (Kucharzik et al 2001). However in contrast to occludin, down-regulation of these other junctional proteins was limited to areas of active inflammation manifested as neutrophil transepithelial migration. Similar findings were reported by Gassler at all (2001) who observed a substantial loss of occludin, ZO1, E-cadherin, and p120- and β-catenins from epithelial junctions in actively inflamed mucosal tissues from UC and CD patients. Interestingly, no changes in expression or distribution of junctional proteins were observed in non-inflamed tissue (Gassler et al 2001). Several conclusions can be drawn from these analyses of intestinal mucosa from IBD patients. First, morphological characteristics of IBD include disassembly of the epithelial AJC and loss of junctional proteins from areas of cell–cell contact. Second, the loss of junctional proteins is only partially determined by their expressional down regulation. Lastly, profound alterations of epithelial junctions are likely secondary to active inflammation and migration of leukocytes.

Proinflammatory cytokines mediate disruption of the AJC and endocytosis of junctional proteins

It is generally accepted that cytokines produced in the intestinal mucosa play an important role in the pathophysiology of IBD. While the relative contributions and mechanisms of action of various cytokines to disease are not well understood, it is clear from *in vitro* studies that proinflammatory cytokines have direct effects on intercellular junctions and permeability. While modulation of epithelial and endothelial tight junction structure and function by cytokines has been recently reviewed elsewhere (Walsh et al 2000), we will focus below on the effects of cytokines on cultured intestinal epithelial cells. Among a variety of proinflammatory cytokines, effects of interferon (IFN)γ and tumour necrosis factor (TNF)α have been the most extensively investigated because increased synthesis of both cytokines is observed in mucosal biopsies from IBD patients (Niessner & Volk 1995). Studies from several laboratories, including ours, have shown that IFNγ increases permeability across T84, Caco-2 and HT-29 intestinal epithelial cell monolayers (Madara & Stafford 1989, Marano et al 1998, Youakim & Ahdieh 1999, Bruewer et al 2003). This permeability increase is manifest by a fall in transepithelial resistance (TER) and increased flux of paracellular markers of different molecular weights. In these *in vitro* studies, TNFα appears to be less potent with no increase in paracellular solute flux and variable decreases in TER depending on which cell line is used (Marano et al 1998, Madara & Stafford 1989, Bruewer et al 2003). However, TNFα significantly potentiates the effects of IFNγ on T84 cells (Bruewer et al 2003). Morphological analyses using freeze-fracture electron microscopy demonstrated that combined treatment of HT-29 cells with IFNγ and TNFα results in a significant decrease in number of strands and complexity of the TJ (Rodriguez et al 1995). These findings resemble alterations of TJs reported in colonic mucosa of IBD patients (Schmitz et al 1999).

The mechanisms underlying cytokine-induced disruption of TJs in intestinal epithelial cells are poorly understood. One possibility that has been proposed involves expressional down-regulation of junctional proteins by cytokines. Indeed, IFNγ has been reported to decrease mRNA and protein levels for ZO1 in T84 cells (Youakim & Ahdieh 1999), whereas TNFα was reported to down-regulate expression of occludin in HT-29 cells (Mankertz et al 2000). Despite these observations, it is not known whether transcriptional down-regulation of junctional proteins is a direct cause of TJ disruption and increased permeability or if down-regulation occurs in response to cytokine-induced disassembly of the AJC. Results of recent studies in our group support the latter possibility (Bruewer et al 2003). As has been shown by others, we observed dramatic increases in paracellular permeability of T84 monolayers exposed to IFNγ alone or in a combination with TNFα. However, extensive biochemical analyses failed to demonstrate reductions in total levels of TJ (occludin, JAM1, claudin 1, claudin 4, ZO1) and AJ (E-cadherin, β-catenin) proteins. Despite the lack of effect on total protein expression, cytokine exposure resulted in dramatic alteration in localization of specific TJ proteins in T84 cells. This redistribution was characterized by discontinuities in lateral membrane staining and sub-membranous accumulation of occludin, JAM1, claudin 1 and claudin 4. In contrast, the distribution of the TJ cytoplasmic plaque protein ZO1 and the AJ proteins, E-cadherin and β-catenin were only minimally affected (Bruewer et al 2003). These observations could be explained by selective internalization of junctional constituents from the plasma membrane or by impaired delivery of newly synthesized proteins from Golgi to intercellular junctions. Further experiments suggested that junctional proteins were indeed internalized since pharmacological inhibition of endocytosis prevented cytokine-induced cytosolic accumulation of TJ proteins, whereas blockage of *de novo* protein synthesis was ineffective. From these observations, we have hypothesized that endocytosis of apical junctional proteins may explain how epithelial junctions become leaky in IBD (Bruewer et al 2003).

Endocytosis of junctional proteins: physiological implications and mechanisms

While the observations above suggest a link between disassembly of junctions/enhanced permeability in IBD and endocytosis, relatively little is known about how intercellular junctions are internalized. Clearly, endocytosis is part of a normal cellular process that regulates the biogenesis of AJs and TJs (reviewed in Ivanov et al 2004). For example, constitutive endocytosis of E-cadherin and/or TJ strands has been shown to take place in kidney epithelial cells, keratinocytes, and adenocarcinoma cells. Likewise, stimulation of epithelial cells with hormones and growth factors induces internalization of AJ (E-cadherin and β-catenin) and TJ (ZO1, occludin and claudin 1) proteins. Tissue remodelling and morphogenesis *in vivo* also involve endocytosis of intercellular junctions. For example, internalization of TJ strands has been observed in the rat intestine and retina, and human fetal hindgut. From such observations, it is clear that endocytosis of intercellular junctions is a necessary part of a number of epithelial cell functions. For example, endocytosis of AJs and TJs allows for rapid down-regulation of cell–cell adhesion that is critical for tissue remodelling, cell migration and extrusion of apoptotic cells. In addition, endocytosis is likely central in repairing a damaged AJC by removing structurally altered junctional proteins from the plasma membrane. Lastly, since junctional proteins have been linked to intracellular signalling events, endocytosis is likely to actively facilitate signal transduction by bringing junction-associated transcriptional regulators into the nucleus.

Internalization of AJs and TJs can be accelerated by a variety of physiological and non-physiological extracellular stimuli. *In vitro* examples of endocytosis of selected junctional proteins in T84 and Caco-2 intestinal epithelial cells in models mimicking an inflammatory insult (IFNγ treatment), epithelial-to-mesenchymal transition (Ca^{2+} depletion), oxidative stress (t-butyl hydroperoxide treatment) and ischaemia (ATP depletion) are shown in Fig. 1. While the time course of endocytosis varies amongst the various stimuli, the end result is nearly complete internalization of the AJC.

As alluded to above, the basic mechanism by which AJs and TJs are endocytosed remains poorly understood. Recently, we began characterizing internalization of AJs and TJs in T84 intestinal epithelial cells using a Ca^{2+}

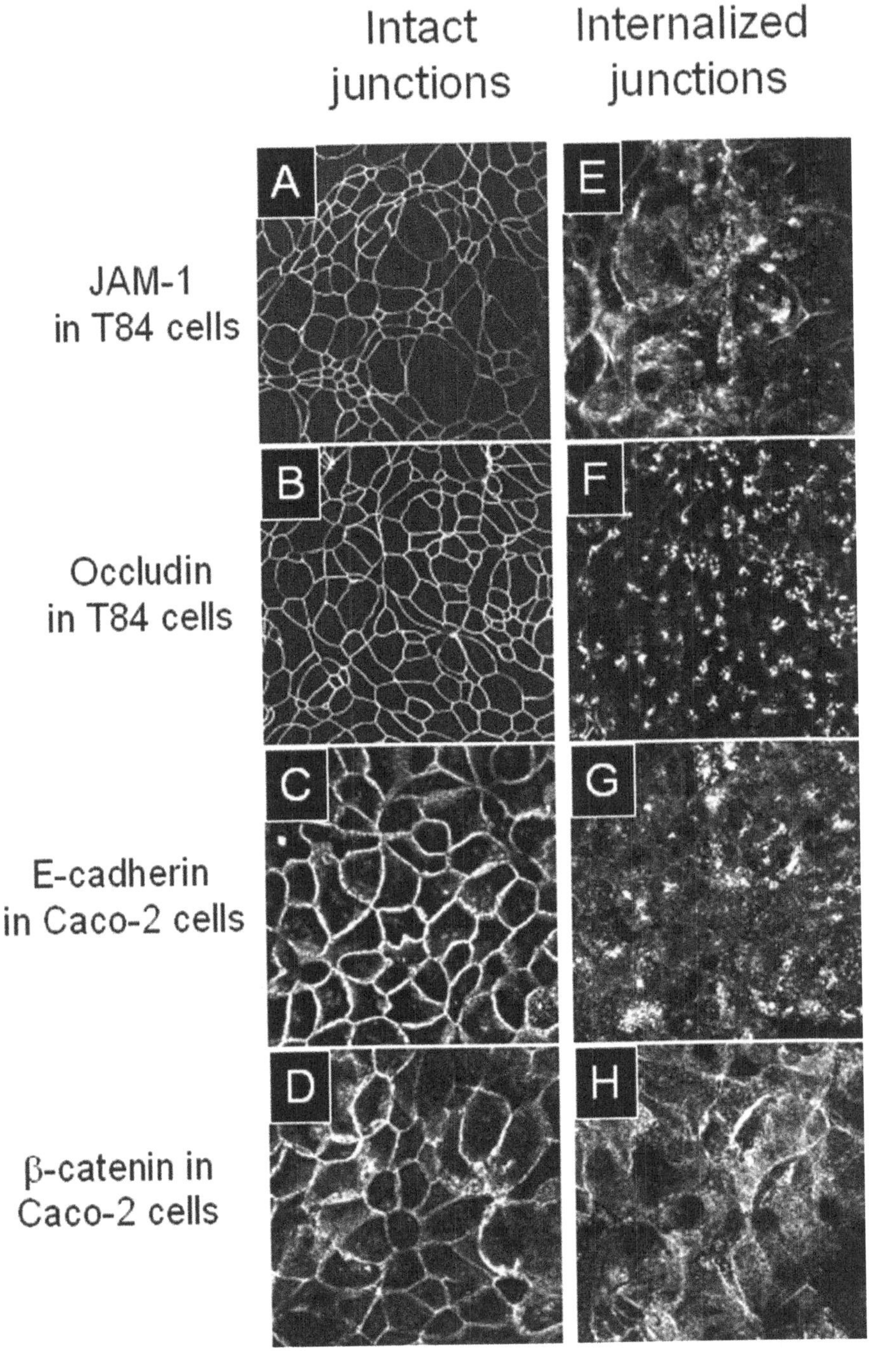
Intact junctions
Internalized junctions
JAM-1 in T84 cells
Occludin in T84 cells
E-cadherin in Caco-2 cells
β-catenin in Caco-2 cells
A
B
C
D
E
F
G
H

depletion model (Ivanov et al 2004). Endocytosis of AJ and TJ proteins in T84 cells after Ca^{2+} chelation appears to be a rapid and highly synchronized event. Double-immunolabelling experiments revealed virtually identical dynamics of internalization for the entire AJC (E-cadherin, β-catenin, p120 catenin, ZO1, JAM1, occludin, and claudin 1). However, despite synchronized endocytosis, there was preserved colocalization of AJ and TJ proteins with their respective junctional members in *different* endosomal populations.

Epithelial cells internalize and sort plasma membrane proteins using specialized endocytotic machineries (Mukherjee et al 1997). The three most extensively characterized pathways include ones using clathrin-coated pits, caveolae and macropinocytosis. From our data, we have concluded that internalization of AJ and TJ proteins in calcium-depleted T84 cells is clathrin-mediated. Indeed, chemical agents preventing the assembly of clathrin-coated pits (hypertonic sucrose, cytosolic acidification and phenylarsine oxide) significantly attenuated endocytosis. Furthermore, both clathrin and its auxiliary protein α-adaptin were recruited to internalized junctional proteins but did not colocalize with intact AJs/TJs. In contrast, we do not observe caveolar-mediated endocytosis since disruption of caveolae fails to prevent internalization of AJs and TJs, and no colocalization of internalized junctional proteins with caveolin 1 was found. Similarly, macropinocytosis is not involved since endocytosis is not inhibited by blocking an obligate mediator of macropinocytosis, phosphatidylinositol-3-kinase, and there is no colocalization of internalized junctional proteins with markers of macropinocytosis (Ivanov et al 2004).

Once internalized, AJ and TJ proteins are delivered to a cytosolic compartment that localizes between the nucleus and the apical plasma membrane. This compartment appears to represent early endosomes since internalized AJ and TJ proteins colocalize with two early endosomal markers, Rab5 and EEA1. However, this association of internalized junctional proteins with early endosomes is brief and disappears at later stages (60–120 min) of calcium depletion. So where do the internalized junctions end up? Surprisingly, endosomes containing junctional proteins do not acquire classical markers of any organelles involved in protein

FIG. 1. Various extracellular stimuli induce endocytosis of apical junctional proteins in intestinal epithelial cell lines. Endocytosis of junctional proteins in confluent T84 (A, B, E, F) and Caco-2 (C, D, G, H) cells was induced by one of the following treatments: incubation with IFNγ (E); depletion in extracellular calcium (F); incubation with t-butyl hydroperoxide (G) or ATP depletion (H). Intracellular localization of TJ proteins JAM1 (A, E) and occludin (B, F) and AJ proteins E-cadherin (C, G) and β-catenin (D, H) was determined by immunofluorescence labelling and confocal microscopy. In control epithelial cells (A–D), all junctional proteins are localized at intercellular contacts revealing a characteristic 'chicken wire' staining pattern. In stimulated cells (E–H), the majority of junctional proteins are internalized and are present either in scattered or aggregated endosomes throughout the cytoplasm.

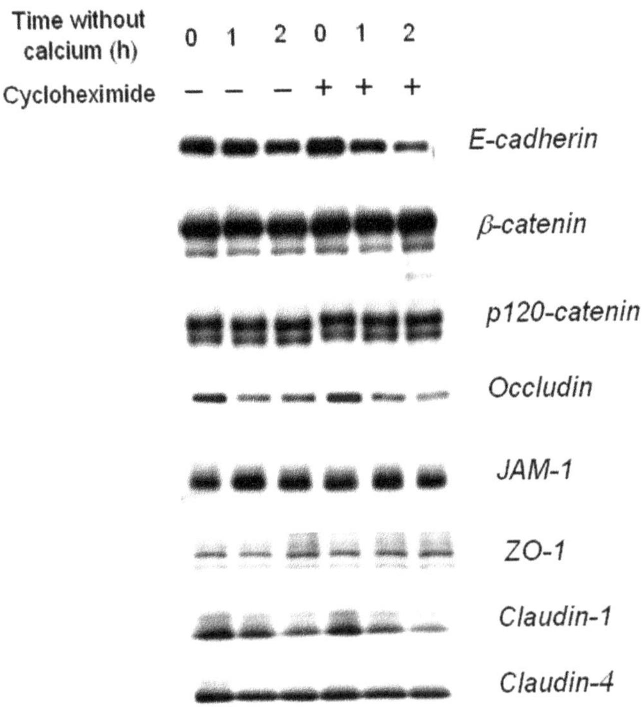

FIG. 2. Calcium depletion results in rapid selective degradation of E-cadherin, occludin and claudin 1 in intestinal epithelial cells. Confluent T4 monolayers were preincubated for 60 min with 20 μM cycloheximide to block protein synthesis or vehicle, followed by 60 or 120 min incubation in calcium-depleted medium containing the same concentration of the inhibitor. Control monolayers were incubated for 120 min in conventional medium with 20 μM cycloheximide or vehicle. After the treatment, the content of different AJ and TJ proteins in total cell lysates was determined by Western blotting. Note that calcium depletion for 60 or 120 min selectively decreases the levels of E-cadherin, occludin and claudin 1 which is accentuated by cycloheximide treatment.

trafficking viz., recycling endosomes, late endosomes/lysosomes, *trans*-Golgi network or the endoplasmic reticulum. These data suggest that at later stages of endocytosis, junctional proteins are delivered into an unusual endosomal compartment distinct from well-characterized intracellular organelles. We find that this late compartment is enriched in the t-SNARE protein, syntaxin 4, and therefore resembles an unusual storage compartment for basolateral membrane proteins previously described in non-polarized kidney epithelial cells (Low et al 2000). Although the function of this syntaxin 4 labelling compartment remains unclear, it may play an important role in determining the fate of internalized proteins. In support of this, we have observed that junctional proteins endocytosed to this compartment have differential stability, with some (E-cadherin, claudin 1 and occludin) being rapidly degraded, while others (β-catenin, p120 catenin, ZO1, JAM1, claudin 4) remain intact at least for several hours (Fig. 2) and can be recycled back to the intercellular junctions. We speculate AJ and TJ proteins internalized to this intestinal epithelial cell compartment are selectively degraded, thus resulting in decreased expression as observed in inflamed mucosa from IBD patients.

In conclusion, recent analysis of IBD tissues together with several model studies on intestinal epithelial cells *in vitro* have provided new insights into mechanisms of barrier disruption in IBD. From these studies, it is clear that disassembly of interepithelial apical junctions is triggered by inflammatory stimuli and results in endocytosis and degradation of both tight and adherens junction proteins. While the process of internalization of epithelial junctions is a necessary part of normal physiology, we predict that acceleration or dysfunction of this process can result in enhanced permeability and the pathophysiological consequences of such. Thus, a better understanding how junctions are internalized may provide ideas for therapeutic modulation of endocytosis to either reduce or enhance intestinal permeability to treat disease or facilitate drug delivery, respectively.

Acknowledgements

Studies from the authors' laboratories were funded by National Institutes of Health Grants DK53202, DK61379, and DK59888.

References

Bruewer M, Luegering A, Kucharzik T et al 2003 Proinflammatory cytokines disrupt epithelial barrier function by apoptosis-independent mechanisms. J Immunol 171:6164–6172

Gassler N, Rohr C, Schneider A et al 2001 Inflammatory bowel disease is associated with changes of enterocytic junctions. Am J Physiol Gastrointest Liver Physiol 281:G216–G228

Hermiston ML, Gordon JI 1995 Inflammatory bowel disease and adenomas in mice expressing a dominant negative N-cadherin. Science 270:1203–1207

Ivanov AI, Nusrat A, Parkos CA 2004 Endocytosis of epithelial apical junctional proteins by a clathrin-mediated pathway into a unique storage compartment. Mol Biol Cell 15:176–188

Jankowski JAZ, Bedford FK, Boulton RA et al 1998 Alterations in classical cadherins associated with progression in ulcerative and Crohn's colitis. Lab Invest 78:1155–1167

Kucharzik T, Walsh SV, Chen J, Parkos CA, Nusrat A 2001 Neutrophil transmigration in inflammatory bowel disease is associated with differential expression of epithelial intercellular junction proteins. Am J Path 159:2001–2009

Low SH, Miura M, Roche PA, Valdez AC, Mostov KE, Weimbs T 2000 Intracellular redirection of plasma membrane trafficking after loss of epithelial cell polarity. Mol Biol Cell 11:3045–3060

Madara JL 1998 Regulation of the movement of solutes across tight junctions. Annu Rev Physiol 60:143–159

Madara JL, Stafford J 1989 Interferon-γ directly affects barrier function in cultured intestinal epithelial cell monolayers. J Clin Invest 83:724–727

Mankertz J, Tavalali S, Schmitz H et al 2000 Expression from the human occludin promoter is affected by tumor necrosis factor α and interferon γ. J Cell Sci 113:2085–2090

Marano CW, Lewis SA, Garulacan LA, Peralta Soler A, Mullin JM 1998 Tumor necrosis factor-α increases sodium and chloride conductance across the tight junctions of Caco-2 BBE, a human intestinal epithelial cell line. J Membr Biol 161:263–274

Meddings JB 1997 Review article: intestinal permeability in Crohn's disease. Aliment Pharmacol Ther 11(suppl 3):47–53

Mukherjee S, Ghosh RN, Maxfield FR 1997 Endocytosis. Physiol Rev 77:759–803

Niessner M, Volk BA 1995 Altered Th1/Th2 cytokine profiles in the intestinal mucosa of patients with inflammatory bowel disease as assessed by quantitative reverse transcribed polymerase chain reaction (RT-PCR). Clin Exp Immunol 101:428–435

Panwala CM, Jones JC, Viney JL 1998 A novel model of inflammatory bowel disease: mice deficient for the multiple drug resistance gene, mdr1a, spontaneously develop colitis. J Immunol 161:5733–5744

Rodriguez P, Heyman M, Candalh C, Blaton MA, Bouchaud C 1995 Tumour necrosis factor-α induces morphological and functional alterations of intestinal HT29 cl.19A cell monolayers. Cytokine 7:441–448

Schmitz H, Barmeyer C, Fromm M et al 1999 Altered tight junction structure contributes to the impaired epithelial barrier function in ulcerative colitis. Gastroenterology 116:301–309

Tibble JA, Bjarnason I 2001 Non-invasive investigation of inflammatory bowel disease. World J Gastroenterol 7:460–465

Tsukita S, Furuse M, Itoh M 2001 Multifunctional strands in tight junctions. Nat Rev Mol Cell Biol 2:285–293

Walsh SV, Hopkins AM, Nusrat A 2000 Modulation of tight junction structure and function by cytokines. Adv Drug Deliv Rev 41:303–313

Yap AS, Brieher WM, Gumbiner BM 1997 Molecular and functional analysis of cadherin-based adherens junctions. Annu Rev Cell Dev Biol 13:119–146

Youakim A, Ahdieh M 1999 Interferon-γ decreases barrier function in T84 cells by reducing ZO-1 level and disrupting apical actin. Am J Physiol Gastrointest Liver Physiol 276: G1279–G1288

DISCUSSION

Powrie: Is it known in molecular terms how IFNγ may induce this type of endocytosis? Is it a direct effect through the receptor and the relevant STATs, or is it a secondary effect?

Parkos: No. A couple of years ago a group reported that this involved changes in regulation of the actin cytoskeleton (Youakim & Ahdieh 1999). The problem with this is that the cytokine-mediated effects are seen at 72 h. We don't even see an effect on the barrier for 24 h. So many things have happened in the cell by that time that it is anybody's guess. It is possible that STATs are involved, but it is difficult to assess at what time to test it and how it directly ties in to the ultimate barrier response.

Powrie: Is there any influence of Toll receptor ligation on this process?

Parkos: This hasn't been looked at. I didn't add the neutrophil-based work that we have done here. We can place a transepithelial gradient of a chemoattractant, and if we stimulate neutrophils to migrate in a basolateral-to-apical direction in a transwell set-up in which the epithelium is on the opposite side of a permeable filter with pores too small for the neutrophils to pass through (and into the epithelium), the permeability still dramatically increases. Phosphorylation occurs at the level of the intracellular junction. We can modulate neutrophil migration pretty dramatically with MALP2, which binds to TLR2 and TLR6, both of which are present on neutrophils. However, TLR2 and TLR4 have also been reported on epithelial cells.

Powrie: It has been suggested that Toll receptor ligation is important for the fusion of certain endocytic compartments in antigen-presenting cells (APCs).

Parkos: Isn't that in Golgi?

Powrie: I'm not sure.

Parkos: There are just a few papers on this. It is localized in the more distal endocytic pathway that we see here. No one has looked at it with regard to tight junctions. Since we don't see any localization in those compartments we would have to go back and see whether TLR2 is there.

Meddings: Early on in your paper you pointed out that in the model you were looking at certain things didn't increase permeability. For example, apoptosis wasn't important. If we go back to physiological systems of bugs interacting with the epithelium, *Giardia* and other microbes do cause apoptosis, and if this is inhibited there is no permeability effect. Bugs also induce other pathways such as the activation of myosin light chain kinase (MLCK). Have you looked at inhibitors of MLCK, and is this related? In the whole scheme of things, is this a big part of the mechanism by which bugs can induce changes in permeability?

Parkos: I could talk for an hour on this! We have tried ML7 to try to get a handle on this. In a paper on Caco-2 cells (Ma et al 2000) it was reported that after 10 min of Ca^{2+} depletion ML7 blocked some of the permeability changes on low-resistance Caco-2 monolayers. We haven't seen an effect of ML7 on inhibition of internalization. If we add ML7 to T84 or Caco cells by itself we also observe a pretty dramatic fall in transepithelial resistance. It may be that activation of the kinase is not involved, but perhaps it is a phosphatase.

Meddings: Do you think this is a significant pathway for altering tight-junctional resistance?

Parkos: We don't know many of the details of the dynamics of tight junctional recycling yet. This is hard to study in natural tissues. If you disrupted this process *in vivo*, it isn't clear what the immediate or long-term effects would be. It is known that E-cadherin turns over in several hours, which is fairly rapid. I don't know whether it is part of the big picture, but it is part of the normal physiology of the intestine. Intestinal epithelial cell extrusion happens over a course of 72 h and acceleration or deceleration of this results in pathological conditions. This is something worth considering. With regard to pathogens, there are reports of *Helicobacter pylori* interacting with ZO1 (Amieva et al 2003). JAM-A was also implicated in this, but I thought that the ZO1 data were more impressive. We know that viruses get into cells by binding to TJ components. It is not clear how lumenal pathogens get access to cells via receptors that are underneath the TJ. It is possible that under certain conditions where you get a little disorganization or recycling of these junction proteins that a small amount redistributes to the apical membrane where interaction with pathogens is possible. Examples include poliovirus and adenovirus that bind to junctional proteins. No one knows what the receptor for rotavirus is but it is not unreasonable to suspect that it may bind to a junctional protein too.

Jewell: For the benefit of the ignorant, can you tell us what MLCK does?

Parkos: MLCK regulates the tension in the perijunctional actomyosin ring. When it is activated there is contraction of the ring and increased paracellular permeability.

Jewell: Do you think the tight junction and the mechanism of the barrier is the same throughout the intestine?

Parkos: The complexity of the tight junction clearly varies in tissues. The most dramatic example of this is with vascular endothelium which contains poorly-formed tight junctions. The blood–brain barrier is an exception which has a resistance of thousands of Ohms and a very complex meshwork of tight-junctional strands. There are clearly permeability differences in the gut and Jim Anderson reported that there are differences in claudin expression here too (Rahner et al 2001). This may account for permeability differences for various solutes. The relative composition of some of these proteins in TJs may be different, even though the actual TJ strand structure is the same. There is controversy over whether occludin is important or not. Occludin knockout mice are viable. Tsukita showed that impedance wasn't all that different, however careful permeability studies on those animals were not performed (Saitou et al 2000). I would suspect that there are important regulatory aspects of barrier function in the guts of those animals that are probably going to be altered by disruption of occludin.

Meddings: It has been shown nicely that the different claudins have different anion and cation selectivity, and their distribution along the gut varies between species. Perhaps most importantly it is different along the length of the crypt–villus axis. Thus immature crypt cells have very different tight junctions from mature villus crypt cells. I think this probably has some implications for disease with rapid turnover and where the death of more mature cells is important.

Sartor: Jim Anderson, using germ-free versus SPF mice has shown that commensal bacteria can up-regulate certain claudins. He looked at about seven, and some were unchanged, some went up and some went down. He has found a difference between the apical surface and the crypt: the ones on the apical surface tended to go up. I have a question. Is the IFNγ/TNF alteration reversible, just as the Ca^{2+} is? Even though the Ca^{2+} is an elegant *in vitro* tool, do you think this has any relevance to the real life situation, particularly in inflammation and the dietary changes that we talked about earlier?

Parkos: The relevance question is frequently raised. With regard to reversibility, if we add in the TNF, it doesn't reverse as well. With IFNγ it does. It takes a while but 72 h after withdrawal of cytokine it increases to about 75% of baseline. Concerning the physiological relevance, we used the Ca^{2+} switch because it is a well established model; everything happens in a rapid and orchestrated fashion. It might be argued that when you take Ca^{2+} away from cells many Ca^{2+}-dependent protein functions will be altered. However, you can do the same experiments by placing epithelial cells in 5 μM Ca^{2+} media. In this case TJs are internalized with the same kinetics and argues that E-cadherin is acting as an extracellular Ca^{2+} sensor. When extracellular Ca^{2+} is reduced below E-cadherin's affinity for Ca^{2+}, then the internalization process begins. With regard to physiologic conditions, we are actively looking at cytokine-induced endocytosis of TJs. This is a little messier to look at because of this dynamic nature of the process—endocytosis doesn't all happen at the same time but rather occurs over a period of many hours. Thus, at any point, there are early endosomes, later endosomes, and so on which complicates colocalization studies. We are finding similarities and differences with the calcium depletion model and thus can't draw definitive conclusions yet.

Schölmerich: We have just looked at Crohn's patients using immunohistochemistry for E-cadherin and β-catenin. It turns out that in the environment of fistulae there is no E-cadherin left in the remaining mucosal layer. The β-catenin left is all intracellular. This is a very interesting observation.

Parkos: You could argue that Ca^{2+} is also important in regulating the epithelial–mesenchymal transformation. Thus, this may also be relevant to tumorigenesis.

Schölmerich: It fits with your endocytosis ideas. β-catenin is not degraded because it remains intracellular, and E-cadherin must be degraded because it is completely gone.

Roediger: As a biochemist I have used Ca^{2+} chelation to isolate cells. I have always believed it was a TJ effect. But there is also ATP depletion in isolated colonocytes in UC. Is the cytokine-effect or ATP-depletion a dominant feature of internalization of these structures?

Parkos: We haven't looked at this in any detail in terms of the kinetics. If we make epithlelial cells anoxic for 30 min, this is what we see. What we need to do is to look at the internalization. It happens quite dramatically. The question is, can we get a more orchestrated internalization with one of these more 'physiological' mechanisms that will make ferreting out the endocytotic pathway easier? This may be the way to go. These data simply show that internalization events can occur after treatment for 30 min.

Kelleher: In terms of the changes in phosphorylation that occur, have you formally looked for changes in serine phosphorylation or tyrosine phosphorylation? I am thinking particularly in terms of the effects of protein kinase C—the Ca^{2+}-dependent forms—which might be modulated by your Ca^{2+} depletion.

Parkos: I haven't looked at PKC with respect to the neutrophil-based effects. We don't see gross disruption of the tight junctions in the course of that assay. We see a dramatic increase in permeability and a fall in transepithelial resistance. We see tyrosine phosphorylation events occurring at the level of the TJ. It doesn't appear to be occludin phosphorylation. We do see MLCK phosphorylation in our model system. There is a protein released by neutrophils called azurocidin, which in the endothelial literature has been shown to alter permeability via MLCK. We suspected that this is what is being released by neutrophils in our epithelial assay thereby activating MLCK and contraction. We tested purified protein from two sources and saw an opposite effect; decreased permeability. Therefore we don't think that the response we observe is secondary to azurocidin. Serine phosphorylation is very interesting in that, with the neutrophil-based experiments we see a granular cytoplasmic staining pattern that moves to the nucleus. This is consistent with translocation of some transcription factor and we are currently setting up the microarrays to look at what happens transcriptionally in epithelial cells exposed to neutrophils. With regard to endocytosis, the only thing I can say is that ML7 doesn't inhibit it in our hands. We think cofilin may be dephosphorylated but haven't looked in detail at protein kinases yet.

Kelleher: In terms of the molecules that do change, have you looked at the $\beta 4$ integrins? I know they are away from the TJs and AJs on the basal surface. Changes in basal cell attachment could cause alterations in the TJ and AJ function by disrupting cell polarity and morphology.

Parkos: In the time-course of our assays the cells stay where they are. If you leave cells in Ca^{2+}-free media overnight, they begin to detach and fall off. Over a 2 h depletion assay, they remain as intact monolayers. We have used fluorescent dyes to look at fence function. During the course of this, we start seeing slight disruption of fence function over a 2 h period. $\beta 1$ integrins are expressed on the basal aspect of the epithelial cells and we have not looked at this area in detail yet.

Moore: You neatly avoided talking about transcellular movement *through* epithelial cells. In the vascular endothelial cell you have transport of infiltrating cells going right through the cell. What is known about epithelial transport?

Parkos: That's a huge topic. There are changes in cyclic nucleotide-mediated secretion of ions that are quite dramatic. If you treat with cytokines you can show that intestinal epithelial cells change from being able to activate chloride secretion (which would be transcellular) to being more of an immune type of cell. In concert with endocytosis of intercellular junctions, with cytokines, at least, there is a dramatic modulation of the ion secretory capacities of these cells.

Moore: But you don't know whether it is possible for a pathogen, for example, to be transported through the cell? Does it have to go through the junction?

Parkos: Some pathogens enter epithelial cells from the apical surface and then go between and presumably across cells. *Listeria* does this, for example. I also mentioned that other pathogens somehow gain access to intercellular junctions as receptors even though these receptors are not expressed on the lumenal surface. *Yersinia* binds to β1-integrin on M cells and intestinal epithelial cells where it is expressed basolaterally. Transcellular movement of cells across epithelia as has been seen in endothelia has not been observed, to my knowledge.

Furrie: I was at a *Clostridium* meeting earlier this year, and found out that *C. difficile* produces a toxin that directly interacts with TJ proteins. Have you any idea about other potential commensal organisms and how they might affect expression and recycling back? If you have a disrupted membrane and you colonize with a probiotic bacterium, can you show increased recycling back to reform the TJ?

Parkos: C. difficile toxA and toxB actually ADP ribosylate Rho-type proteins. These are intimately involved in the cytoskeletal regulation of TJs. Asma Nusrat showed a couple of years ago that you can get movement of occludin into caveolar-like structures in this system, so there is very dramatic disruption of the TJ. Others have shown that these toxins can be directly chemotactic towards neutrophils. However, it is not clear what relative contributions the following have to the pathophysiology of disease: induction of chemotaxis, toxicity to the cells, or junction disruption via direct interactions with TJ proteins. Another toxin, CNF from *E. coli* activates Rho kinase, which then modulates the actin cytoskeleton to cause dramatic effects on the barrier. *C. perfringins* toxin binds claudin 4, and it does so by getting into the cell and binds to the cytoplasmic tail.

Fedorak: What about the mycoplasma that Bill Roediger told us about?

Parkos: I like that; it is a great idea.

Mahida: Just to add to your list, *B. fragilis* toxin cleaves the extracellular portion of E-cadherin, leading to reversible loss of barrier function. We have found that with low concentrations of *C. difficile* toxin A it is possible to reverse the loss of barrier function, which can be accelerated by transforming growth factor (TGF)β. There is some interest in cytokines that may enhance barrier function. Have you looked at these in your system?

Parkos: No. I remember this work well, though. If you think of E-cadherin in terms of its short cycling time, then if it was selectively clipped with a toxin you wouldn't kill the cell but this would be a reversible process. In contrast, if you saturate all the Rho proteins with an irreversible modification, this might be a bit harder to recover from.

Ghosh: You mentioned the *MDR1* knockouts. What was the nature of the cross-talk between ABC transporters and the TJ/AJ proteins? Do they talk to each other?

Parkos: I don't know. They have a pretty clear permeability defect. What is interesting about this model is that the cadherin dominant negative report several years ago showed a permeability defect and colitis. In this case the animals all get colitis that apparently can be modulated by antibiotics.

Rhodes: What does low dose detergent do to tight junctions?

Parkos: This is a good way to sort out what we call Triton-soluble versus -insoluble pools. Junctional adhesion molecule is an interesting protein in that 5% of it is triton insoluble and 95% is soluble. In contrast, almost all of ZO1 is Triton insoluble, and this is thought to be the scaffolding protein that is intimately associated with the TJ and the actin cytoskeleton. For JAM-A, you can treat monolayers with 0.5 Triton X-100 and get rid of some of the confounding lateral staining and non-TJ-associated staining. With some of these experimental manipulations you can show changes at a morphological level due to the experimental manipulations: associations that you wouldn't be able to show on gels. In short, a big part of the TJ complex is classically described as being detergent-insoluble in Triton.

Rhodes: What would happen to barrier function and electrical resistance with low concentrations of Triton?

Parkos: You can selectively and carefully punch holes in the apical membrane. If you punch holes just in the apical membrane the barrier is maintained, because the lateral and basolateral membrane are intact. This has been used to introduce compounds into cells. If you punch holes laterally, I would think you would also be able to maintain the barrier just as long as you don't punch holes in both the apical and basolateral membranes.

Bjarnason: For the benefit of the people who do think that the intestinal bacteria are doing some good, could you tell them why you don't sprinkle your experimental system with intestinal bacteria? Presumably the epithelial cells would die.

Parkos: Yes, I think the T84 cells would eventually die from bacterial overgrowth.

Bjarnason: Even if you put these good bacteria on the cells?

Parkos: They are being cultured and the bacteria would overgrow them.

Sartor: In monolayers, if you add bacteroides, for example, you will induce NF-κB activation through TLR4. *E. coli* and LPS will do the same thing. If, however, you co-culture lamina propria and circulating cells, you will interfere with that NF-κB interaction. There is cross-talk between the epithelial cell and the lamina propria—if you can believe these reductionist systems—that would decrease the ability of the epithelial cell to respond to these bacterial stimuli. It is not as simple as one cell doing something in isolation: there is bi-directional cross-talk.

Mahida: Also, probiotics have been shown to have a beneficial effect in monolayers in terms of TJ function.

Parkos: Are you asking from the standpoint of just letting the bacteria grow?

Bjarnason: Yes.

Parkos: In the colon there is a defined concentration of bugs, and the crypts are sterile. Balfour Sartor's point is well taken that under controlled conditions you can put a lot of bacteria on the apical surface of these co-culture models and measure these effects, but if you let them grow unabated they will consume all of the growth nutrients and the epithelial cells will die.

Schölmerich: If you take mice that are deficient for TLR9 and induce colitis, there will be much less inflammation and the normal cytokine increase isn't seen. This is because the bacterial DNA normally stimulates through TLR9. But you see far more defects in the epithelium because the bacteria and their input via TLR9 are needed for inducing the repair function. There is a nice story: if you think about a hippo, it sits in water and defecates into the water. It has a short tail like a propeller which swirls its own faeces over its body. If the tail is cut off the hippo gets skin disorders. The same experiment can be done with frogs. They need their skin to be covered with their faeces from time to time, and they do this by pumping when they sit in the water.

Meddings: John Rioux's comment about solubilization brought up another thought. That is, when we use detergent to solubilize several things happen: membrane structure is altered, for example. One of the things that is sometimes overlooked is that protein function within membranes is also very dependent on the membrane's physical structure and how rigid or fluid the membranes are. A few years ago with Phil Sherman we did an interesting experiment looking at enteroadherent *E. coli* and how they adhere to HEP2 cells. If we slightly fluidize HEP2 cells with a compound called A_2C, a mild detergent, this doesn't change the cholesterol–phospholipid ratio, but enteroadherent *E. coli* would no longer bind. Whatever the receptor was binding before was presumably still there, but its function was such that it couldn't recognize it.

Schölmerich: It might have changed the raft functions.

Meddings: That is possible, although these things will change the non-raft portions of the membrane far more easily than they will change the raft.

References

Amieva MR, Vogelmann R, Covacci A, Tompkins LS, Nelson WJ, Falkow S 2003 Disruption of the epithelial apical–junctional complex by Helicobacter pylori CagA. Science 300:1430–1434

Ma TY, Tran D, Hoa N, Nguyen D, Merryfield M, Tarnawski A 2000 Mechanism of extracellular calcium regulation of intestinal epithelial tight junction permeability: role of cytoskeletal involvement. Microsc Res Tech 51:156–168

Rahner C, Mitic LL, Anderson JM 2001 Heterogeneity in expression and subcellular localization of claudins 2, 3, 4, and 5 in the rat liver, pancreas, and gut. Gastroenterology 120:411–422

Saitou M, Furuse M, Sasaki H et al 2000 Complex phenotype of mice lacking occludin, a component of tight junction strands. Mol Biol Cell 11:4131–4142

Youakim A, Ahdieh M 1999 Interferon-γ decreases barrier function in T84 cells by reducing ZO-1 level and disrupting apical actin. Am J Physiol Gastrointest Liver Physiol 276: G1279–G1288

Apoptosis or necrosis—colonic epithelial cell survival

Peter R. Gibson

Department of Gastroenterology, Box Hill Hospital, Box Hill, Victoria, Australia

Abstract. The organization of the colonic epithelium is directed towards maintaining a continuous layer of cells with functional maturity at the surface, with a constant supply of epithelial cells of sufficient maturity from the crypts. Cells die by shedding at the surface or by undergoing apoptosis *in situ*, followed by shedding or phagocytosis. The nature of the predominant form of cell death in the normal colon remains uncertain but probably involves more than one route. Death of abnormal cells in the crypt is an important process to prevent clonal expansion and tumour formation. Apoptosis of normal cells is largely secondary to the effects of exposure to luminal factors, such as short-chain fatty acids and surfactants; loss of cell–substratum adhesion (anoikis) in the surface compartment; or cytokine-, oxidant stress- or lymphocyte-mediated death in association with inflammation. In inflammatory bowel disease (IBD) cell death can directly or indirectly compromise barrier function, impair effective epithelial restitution, lead to loss of stem cells and regenerative capacity, and has relevance to carcinogenesis. Therapeutic targeting of cell death in IBD is a double-edged sword—promotion of cell survival may favour more effective healing for active inflammation, while promotion of apoptosis may be protective from carcinogenesis.

2004 Inflammatory bowel disease—crossroads of microbes, epithelium and immune systems. Wiley, Chichester (Novartis Foundation Symposium 263) p 133–150

The colonic epithelium is a complex structure made up of cells that are widely heterogeneous by virtue of their position within the epithelial hierarchy or of their state or pathway of differentiation. Understanding the birth, maturation and death of epithelial cells is a key element to understanding the function of the colonic epithelium and its role in disease pathogenesis. This review addresses the issue of cell death within the colonic epithelium, how it occurs in health and disease, and the implications it might have in the pathogenesis and treatment of inflammatory bowel disease (IBD).

The overall organization of the colonic epithelium

The organization of the colonic epithelium is directed towards maintaining a continuous layer of cells with functional maturity at the surface, where the controlled internal environment of the body interfaces with the external world. To achieve this, epithelial cells are born and mature in the crypts, which provide an environment relatively protected from the hostile lumen. This protection derives from the constant, one-way flow of mucus out of the crypt preventing access of the crypt to hazardous soluble and microbial substances within the lumen, and from the low bacterial count within the crypt lumen by virtue, at least in part, of the secretion of defensins and other antibacterial substances. The needs of the surface compartment in terms of cell number and the cellular output of the crypt compartment appear to be closely co-ordinated. For example, the rate of proliferation in the crypt is elevated when excessive cell injury occurs at the surface, and is depressed when that injury is prevented. The mechanisms by which cells in the basal parts of the crypt respond to surface events and needs have not been elucidated.

Modes of cell death in the colon

The fate of epithelial cells in the normal colon is a controversial issue with wide variations in the concepts presented. Reasons for such diverse beliefs include methodological difficulties and artefacts, not factoring in the importance of prevailing microenvironmental conditions, and the fact that there are multiple ways of dying. A further complicating issue is that goblet cells and absorptive cells (often referred to as colonocytes) have different patterns of behaviour. Goblet cells have a longer lifespan, but their fate is unknown. They seem to disappear at the crypt neck. Ultrastructurally, 'dark' cells in the surface epithelium might represent goblet cells depleted of mucus (Barkla & Gibson 1999), but this is far from clear. Absorptive cells will be discussed in the remainder of this review.

There are two possibilities for the fate of epithelial cells in the colon.

Viable cells are shed into the lumen

The general concept is that epithelial cells adhere more loosely to the substratum in the surface compartment than they do in the crypt, and that shear forces exerted on those cells by colonic contents during normal colonic motility dislodge them while still viable. This is the traditional view and derived from observations in the small intestine (Shibahara et al 1995). There is little doubt that cells at the surface are loosely attached to the basement membrane, as seen on many tissue sections (Fig. 1A). Mechanisms for the loose adherence have been proposed. For example, urokinase is secreted by colonic epithelial cells, which express urokinase receptors on their basolateral surface (Gibson et al 1994, Gibson & Rosella 1996). Local activation of proteases would effectively loosen cell–substratum adhesion.

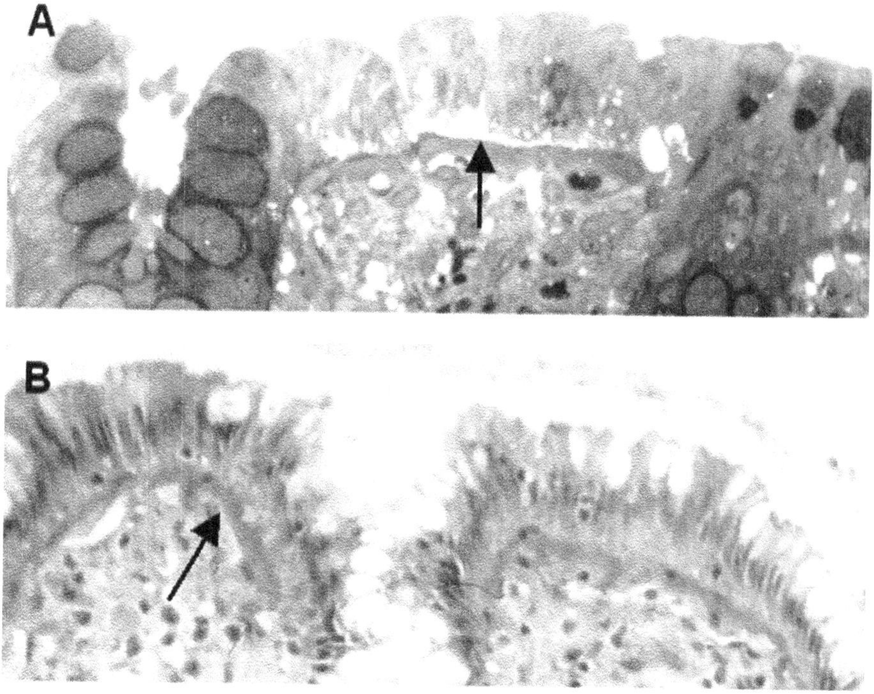

FIG. 1. Loose cell–substratum attachment in the surface compartment of the colonic epithelium. (A) Toluidine blue-stained section showing epithelium loosely attached to the basement membrane (arrow). (B) Immunoperoxidase staining using an anti-urokinase antibody, showing specific staining of the basement membrane and basal membrane of the epithelial cells. Control sections showed no non-specific staining (not shown).

Expression of urokinase is greater at the surface compartment (Fig. 1B) and in the villus tip of the small intestine (Gibson et al 1998).

The best evidence that cell shedding does occur has come from rat studies, where epithelial DNA is readily detected in the faeces (Van Lieshout et al 2004). In humans, long fragments of DNA are not detectable in the faeces of healthy subjects (Boynton et al 2003). These data are best interpreted as showing that cell shedding is a real phenomenon, but is not a major cause of cell loss in association with the hostile luminal conditions on the human large bowel.

Cells die in situ

The general concept is that epithelial cells die while still a part of the epithelium. Two distinct forms of cell death are described:

Necrosis. This refers generally to the phenomenon of 'explosion'—cell swelling, the collapse of the plasma membrane and then cell lysis—and is predominantly a passive phenomenon, a result of cell injury. The release of intracellular contents potentially leads to inflammatory responses. In the strict pathological sense, the term, 'necrosis' refers to the tissue changes secondary to cell death, not the cell death itself.

Apoptosis. This is a process of 'implosion'—the cell first shrinks and then the nucleus condenses, and the cell then disintegrates into apoptotic bodies—and involves gene expression. Since apoptotic cells are shed to the lumen or phagocytosed before disruption of cell membranes occurs, apoptosis is not associated with induction of inflammatory responses due to the release of cellular contents.

This classification is an oversimplification since features used to distinguish the two have been observed in the same dying cells, and is likely to represent two ends of a spectrum (Proskuryakov et al 2003). Nevertheless, it remains a useful conceptual classification since it is definable morphologically and may give clues as to the pathogenic events, both leading up to and as a consequence of cell death.

Reasons for cell death *in situ*

There are several potential reasons why cells in the colonic epithelium die. The first is that, with ongoing cell proliferation and the continued migration of cells from crypt base to surface, there is no room for cells at the surface. Those more loosely bound to the substratum will be pushed off. This is the classical view of cell shedding. The second classical view is that cells are pre-programmed to differentiate and die by apoptosis. Whether programmed cell death and differentiation are linked, and an inevitable sequence in intestinal epithelial cells, is uncertain.

Cells also die because they are abnormal. With such a high rate of replication and with the hostile pro-carcinogenic conditions that characterize the colon, DNA errors during mitosis will not be unusual. Such cells self destruct as an appropriate protective response, usually within the crypt. This is purported to occur via apoptosis but apoptotic cells within the crypt are rarely seen in the normal colon or in familial adenomatous polyposis (Barkla & Gibson 1999, Strater et al 1995). In contrast, cells in the lower third of the crypts were not infrequently observed to have features of necrosis, a hitherto undescribed feature (Fig. 2). The significance of this finding and relationship to abnormal epithelial cells remain unknown.

A major reason for cell death in the colon is from the effect of extraneous factors. Luminal surfactants, such as bile acids and haem, are associated with increased

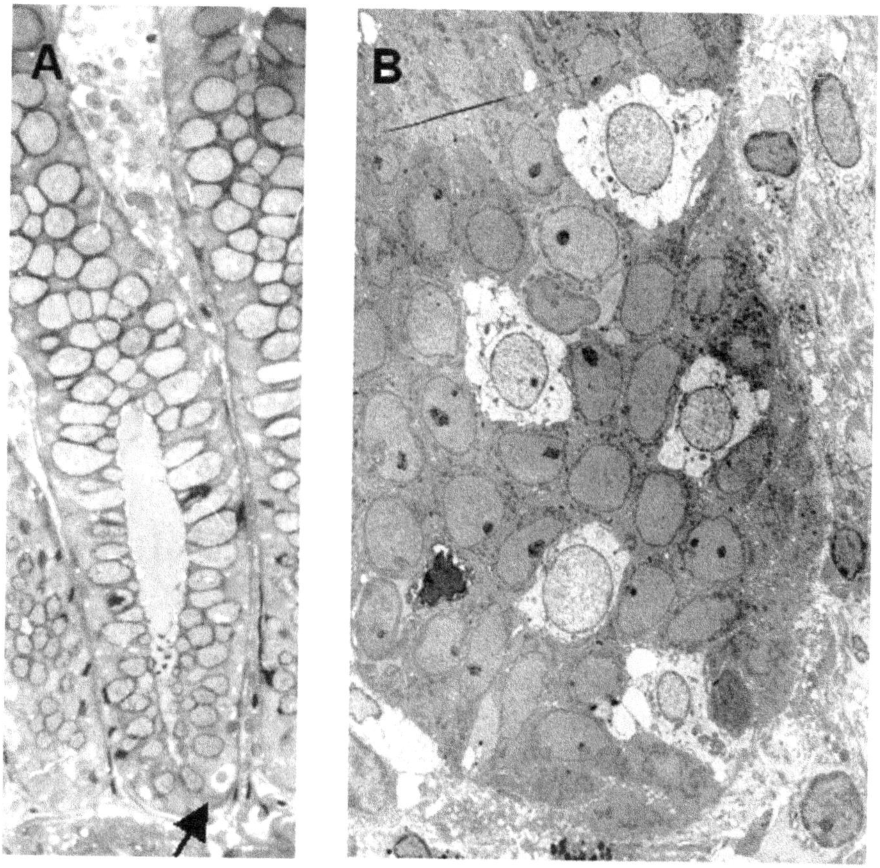

FIG. 2. 'Clear' cells showing features of necrosis in the basal area of colonic crypts. (A) Toluidine blue-stained section with necrotic cell indicated by the arrow. (B) Transmission electron micrograph from a different patient showing multiple necrotic cells. (Reproduced with permission from Barkla & Gibson 1999).

epithelial cell turnover (Lapre et al 1993, Sesink et al 1999). This is likely to be secondary to plasma membrane injury, although cell necrosis that might be anticipated is not observed histologically. Colonic epithelial cells are subject to several pro-apoptotic influences. Loss of cell-substratum adhesion, which is more likely to occur in the surface compartment (as above), leads to apoptosis in colonic epithelial cells, a phenomenon termed anoikis (Strater et al 1996). In inflammatory states, cell-death receptor and mitochondrial pathways for apoptosis can be activated by, for example, tumour necrosis factor (TNF)α or oxidant stress, while cytotoxic lymphocytes can induce apoptosis by lethal delivery of perforans or granzyme A and B to the epithelial cells. In normal colon, lymphocytes with cytolytic granules do not enter the intraepithelial compartment, but in microscopic colitis they do (see below). The surface epithelium is normally exposed to high concentrations of short-chain fatty acids, particularly butyrate, which are powerful inhibitors of the histone deacetylase (HDAC) family of enzymes. One end-result of this is the induction of apoptosis. It is somewhat surprising that the epithelial cells can survive at all in the presence of butyrate at luminal concentrations (often 20 mM or greater). Cells in the crypt are protected from such high concentrations by mechanisms outlined above. 'Escape' mechanisms that protect against butyrate-mediated toxicity are likely to be part of the maturation process. Recent studies in the Caco-2 cell line model of the colonic epithelium have been instructive on how the cells protect themselves. Differentiation of Caco-2 cells in confluent culture has many similarities to the spontaneous maturation that occurs in cells as they migrate from crypt base to neck (Mariadason et al 2000). During that process, Caco-2 cells develop more efficient β-oxidation (more effectively protecting the intracellular environment from high butyrate concentrations) and become resistant to the pro-apoptotic effect of HDAC inhibitors by unknown mechanisms (Mariadason et al 2001).

Patterns of cell death in the normal colon

Apart from our own observations of cell necrosis in the basal third of normal crypts (Barkla & Gibson 1999), features of necrosis are rarely if ever observed in normal colon. However, there is now a large literature on apoptosis in the colonic epithelium. In order to interpret published data, two issues must be understood. First, the morphological features of apoptosis or detection of internucleosomal DNA cleavage using, for example, the TUNEL technique are late features of apoptosis and are estimated to be present for less than 60 minutes. Thus, the rate at which apoptosis is occurring will be much greater than might initially be considered from the absolute number of apoptotic cells observed. Secondly, apoptosis in the surface epithelium of the colon can be induced within minutes of removal of the tissue from its blood supply (Hall et al 1994). Any histological evaluation of apoptosis in the colon, therefore, must be made in tissue that is rapidly fixed or treated in some way to inhibit this event. As an example of this problem, a recent study examined cells early in apoptosis (positively staining for activated caspase), prior to DNA changes, and concluded that apoptosis is initiated as cells enter the surface compartment prior to shedding (Grossman et al 2002). However, since the investigators used resected colon without attention to this potential artefact, their conclusion is also flawed and not valid. Where careful attention is paid to avoiding artefacts by rapid fixation and minimal non-traumatic handling of the tissue, apoptotic cells are found uncommonly at the surface and rarely in the crypts (Barkla & Gibson 1999), even when a marker of

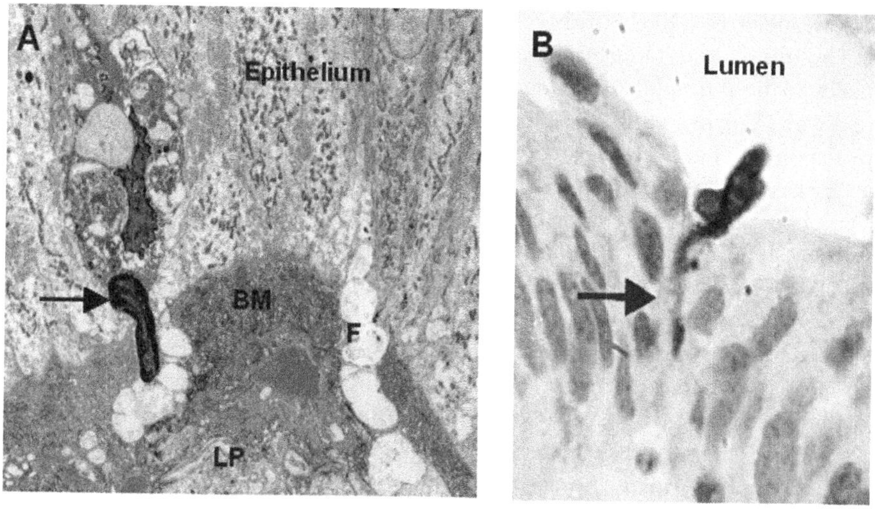

FIG. 3. Fate of apoptotic cells in the surface compartment of the colonic epithelium. (A) Transmission electron micrograph from a normal colon showing an apoptotic nucleus (arrow) within a fenestra (F) in the basement membrane (BM), interpreted as an apoptotic cell passing into the lamina propria (LP) to be phagocytosed. An additional apoptotic cell in seen above the arrowed apoptotic nucleus. (Reproduced with permission from Barkla & Gibson 1999.) (B) Immunoperoxidase staining using the M30 Cytodeath antibody of a section of colonic mucosa from a patient with microscopic colitis. The thin residual cytoplasm of an apoptotic cell (arrow) is seen between viable epithelial cells with the majority of the cell being extruded into the lumen.

very early apoptosis (M30 Cytodeath antibody) is used (P. P. Tagkalidis, P. S. Bhathal, P. R. Gibson, unpublished observations).

Apoptotic cells will either be shed to the lumen (Fig. 3B) or be phagocytosed. In the guinea-pig small intestine, macrophages enter the epithelium and phagocytose dying epithelial cells (Iwanaga et al 1993). In humans, macrophages are not seen within the epithelial compartment, yet subepithelial macrophages have a cellular content consistent with ongoing phagocytosis of apoptotic cells. Furthermore, in patients with melanosis coli, macrophages contain pigment derived from epithelial cells, suggesting that somehow they gain access to epithelial remnants. Based upon ultrastructural observations (Fig. 3A), we have suggested that at least some contracted apoptotic epithelial cells are washed through fenestrae in the subepithelial basement membrane via the current of absorbed water with subsequent phagocytosis by macrophages (Barkla & Gibson 1999).

Epithelial cell death in IBD

Most epithelial cell death in ulcerative colitis and Crohn's disease follows an apoptotic pattern (Iwamoto et al 1996, Kruidenier et al 2003), although in severe disease, necrosis is likely to also occur. Focal areas of necrosis have been described in Crohn's disease (Dourmashkin et al 1983). Mechanisms for such apoptosis are likely to be multiple, including cytokine-activation of the cell death receptor pathway (Yukawa et al 2002), oxidant stress (Kruidenier et al 2003) and other pro-apoptotic factors associated with the mucosal inflammatory response.

We have recently characterized epithelial cell death in patients with microscopic (lymphocytic and collagenous) colitis (P. P. Tagkalidis, P. S. Bhathal, P. R. Gibson, unpublished observations). Colonic epithelial turnover was markedly increased with proliferating cells often seen in the surface compartment. Cell necrosis was not observed but apoptosis of epithelial cells mainly confined to the surface compartment was prominent. Morphological appearances suggested that apoptosis was occurring *in situ* and that the dying cells were subsequently shed to the lumen or phagocytosed subepithelially. Mechanisms for epithelial apoptosis were also examined. A characteristic and diagnostic feature of this condition is a marked increase (usually more than 10-fold) in intraepithelial lymphocytes (IELs). About 15% of IELs possessed cytolytic granules expressing granzyme B, compared with their absence in normal colon. These cells presumably played a role in inducing apoptosis in epithelial cells, although they were evenly spread throughout the epithelium without lying in close proximity to apoptotic cells. The cytokine profile of the mucosa, in which there was a predominance of interferon (IFN)γ, TNFα and interleukin (IL)15, may have contributed to the apoptosis of epithelial cells, although this was not directly studied.

Relevance of epithelial cell death to IBD

Epithelial cell death and its rate are of relevance to three aspects of IBD.

Relevance to the pathogenesis of mucosal inflammation

Changes of epithelial turnover have the potential to alter epithelial differentiation and function. One aspect of great relevance to IBD is barrier function. Animal studies have taught us that impairment of any component of the mucosal barrier may result in mucosal inflammation. Furthermore, accumulating evidence suggests that abnormal epithelial permeability might be an important precursor of Crohn's disease.

Epithelial cell death can potentially compromise barrier function in two ways:

Indirectly by leading to a shorter life span that does not permit appropriate cell differentiation prior to reaching the surface epithelium. Though there are few data available on the effect of changes in epithelial turnover in the non-inflamed colon on barrier function, it is clear that dietary changes in epithelial turnover can alter barrier function. Rats fed a diet containing no fibre have a leaky distal colon, an effect likely to be due to the relative atrophy such a diet causes and the subsequent reduced differentiation of the epithelial cells (Mariadason et al 1999). Feeding rats wheat bran or methylcellulose leads to epithelial growth and greater expression of markers of differentiation, together with a considerable improvement in barrier function (Mariadason et al 1999). We have previously shown that rats ingesting diets containing an excess of highly fermentable fibre (such as resistant starch or guar gum) develop elevated epithelial turnover and injury in the distal colon, but, while the diets promoted carcinogenesis, their effect on barrier function was not studied (McIntyre et al 1993, Young et al 1996, Gibson et al 1999a). This has been recently clarified. In rats fed fructo-oligosaccharide, a highly fermentable and non-absorbed carbohydrate, both paracellular permeability and the rate of colonic epithelial proliferation increased, together with increased susceptibility to *Salmonella* infection (Bovee-Oudenhoven et al 2003, R. van der Meer, personal communication). Whether these effects occur in humans has not been directly examined, but diversion of the faecal stream from colonic mucosa leads to epithelial atrophy and often colitis.

In ulcerative and collagenous colitis, colonic permeability is increased in the presence of mucosal inflammation. Structural abnormalities of the tight junctions (reduced strand count) in severe ulcerative colitis (Schmitz et al 1999) and down-regulation of the expression of tight junction molecules in collagenous colitis (Burgel et al 2002) have been documented. In milder inflammation in patients with ulcerative colitis, however, electrophysiological imaging has indicated that permeability changes reflected focal leaks (see below) rather than tight junction dysfunction.

Directly by causing focal leaks in the barrier. Since cell death at the surface compartment is a normal process, it might be anticipated that mechanisms are in place to protect the barrier when a cell dies by shedding or *in situ* apoptosis. Indeed, morphological studies in the small intestine and in cell lines indicate that a 'zipper' mechanism exists, maintaining continuity of cells (Madara 1990, Abreu et al 2000). Our unpublished observations in microscopic colitis suggest that a similar mechanism occurs in the colon. Induction of apoptosis in colonic epithelial cell lines does increase paracellular permeability, although the degree of change is surprisingly modest (Abreu et al 2000). Application of a conductance scanning technique together with histology on resected colon has indicated the quantitative importance of physical disruption of the continuity of the epithelium, such as the presence of ulceration, erosions, or just nests of apoptotic cells, in compromising barrier function in ulcerative colitis (Gitter et al 2001).

Relevance to epithelial healing

The continued loss of cells from a depleted epithelial pool will have obvious detrimental effects on epithelial restitution and regeneration. Severe inflammation can also be associated with total crypt destruction and the death of stem cells at the crypt base. This can be at least partly dealt with by regenerating crypts via fission, but severe loss will reduce the source of new epithelial cells, compromise the ability to heal, and result in atrophic mucosa if healing is achieved.

Relevance to carcinogenesis

Failure of cells with abnormal DNA to undergo apoptosis is an important pathogenic step in the establishment of clones of mutated cells. The rate of cell death at the surface is a major determinant of the rate of colonic epithelial proliferation. The higher the proliferation, the greater the chance will be that critical mutations occur. Considerable attention has been paid to both of these aspects in the area of cancer prevention (reviewed by Johnson 2002, Ricchi et al 2003).

Therapeutic targeting of cell death

In patients with IBD, therapeutic targeting of cell death in the prevention of relapse or development of carcinoma is a double-edged sword. On the one hand, promotion of apoptosis may have a protective effect in carcinogenesis, but this might potentially compromise barrier function by increasing the rate of cell death with possible generalized reduction of barrier function and increase in focal leaks. Non-steroidal anti-inflammatory drugs (NSAIDs) provide the immediate example of such a dilemma — they are protective of carcinogenesis via promoting apoptosis (Ricchi et al 2003), but compromise barrier function with an associated increased risk of precipitating relapse of IBD.

Different considerations must apply if cell death is being targeted for inducing remission. Inhibition of cell death — that is, promotion of cell survival — would clearly be the goal, so that maximal benefit from epithelial regeneration and restitution can be achieved. Manipulation of key intracellular cell-survival signalling pathways would be the target, underlining the importance of defining the cell death pathways that are actually activated in IBD. Information regarding external cell-survival factors that might be utilized is limited, but relevant factors could include anti-apoptotic inflammatory mediators such as leukotrienes (Ohd et al 2000), stimulators of COX-2 expression, or nutrients such as glutamine, which prevents cytokine-induced apoptosis in HT29 cells (Evans et al 2003). The choice of agents or dietary manipulations that will improve the health of the epithelium must take into consideration the fact that regenerating and 'activated' epithelium may respond differently to normal epithelium (Gibson et al 1999b). Not only is the epithelium different in its state of differentiation but the exposure of more sensitive cryptal epithelial cells to luminal substances will be different. For example, high butyrate concentrations in the lumen are considered appropriate for the health of the colonic epithelium, promoting maturation and good barrier function, suppressing carcinogenesis, and suppressing inflammatory processes. However, in ulcerative colitis, colonic epithelial cells are poor at oxidising butyrate, crypt cells are likely to be exposed to higher luminal butyrate concentrations because of distortion of crypt architecture and reduced mucus output, and the cells will not have had the time to mature to develop protective mechanisms against butyrate-mediated apoptosis. It is not surprising that butyrate enemas have not met with much success as a remission-inducing therapy.

Conclusions

The study of epithelial cell death in the colon, especially its mechanisms and promotion, is currently a hotly pursued area in cancer research. In contrast, our understanding of the functional significance, especially with regard to barrier function, of changes in rates and patterns of cell death is limited and receives scant attention. Since minimization of cell death should be the preferred goal in the treatment and prevention of relapse in IBD, a thorough understanding of the complexities of the colonic epithelium in health and disease should underscore any attempted therapeutic manipulation of colonic epithelial turnover, so that more rational choices can be made.

References

Abreu MT, Palladino AA, Arnold ET, Kwon RS, McRoberts JA 2000 Modulation of barrier function during Fas-mediated apoptosis in human intestinal epithelial cells. Gastroenterology 119:1524–1536

Barkla DH, Gibson PR 1999 The fate of epithelial cells in the human large intestine. Pathology 31:230–238

Bovee-Oudenhoven IM, Ten Bruggencate SJ, Lettink-Wissink ML, Van Der Meer R 2003 Dietary fructo-oligosaccharides and lactulose inhibit intestinal colonisation but stimulate translocation of salmonella in rats. Gut 52:1572–1578

Boynton KA, Summerhayes IC, Ahlquist DA, Shuber AP 2003 DNA integrity as a potential marker for stool-based detection of colorectal cancer. Clin Chem 49:1058–1065

Burgel N, Bojarski C, Mankertz J, Zeitz M, Fromm M, Schulske JD 2002 Mechanisms of diarrhea in collagenous colitis. Gastroenterology 123:433–443

Dourmashkin RR, Davies H, Wells C et al 1983 Epithelial patchy necrosis in Crohn's disease. Hum Pathol 14:643–648

Evans ME, Jones DP, Ziegler TR 2003 Glutamine prevents cytokine-induced apoptosis in human colonic epithelial cells. J Nutr 133:3065–3071

Gibson PR, Rosella O 1996 Abnormalities of the urokinase system in colonic crypt cells from patients with ulcerative colitis. Inflamm Bowel Dis 2:105–114

Gibson PR, Rosella O, Rosella G, Young GP 1994 Constitutive secretion by colonic epithelium of urokinase-type plasminogen activator and plasminogen activator inhibitor-1. Gut 35:969–975

Gibson PR, Birchall I, Rosella O et al 1998 Urokinase and the intestinal mucosa: evidence for a role in epithelial cell turnover. Gut 43:656–663

Gibson PR, Rosella O, Wilson AJ et al 1999a Colonic epithelial cell activation and the paradoxical effects of butyrate. Carcinogenesis 20:539–544

Gibson PR, Nov R, Fielding M et al 1999b The relationship of hydrolase activities to epithelial cell turnover in distal colonic mucosa of normal rats. J Gastroenterol Hepatol 14:866–872

Gitter AH, Wullstein F, Fromm M, Schulzke JD 2001 Epithelial barrier defects in ulcerative colitis: characterization and quantification by electrophysiological imaging. Gastroenterology 121:1320–1328

Grossman J, Walther K, Artinger M, Rummele, P, Woenckhaus M, Scholmerich J 2002 Induction of apoptosis before shedding of human intestinal epithelial cells. Am J Gastroenterol 97:1421–1428

Hall PA, Coates PJ, Ansari B, Hopwood D 1994 Regulation of cell number in the mammalian gastrointestinal tract: the importance of apoptosis. J Cell Sci 107:3569–3577

Iwamoto M, Koji T, Makiyama K, Kobayashi N, Nakane PK 1996 Apoptosis of crypt epithelial cells in ulcerative colitis. J Pathol 180:152–159

Iwanaga T, Han H, Adachi K, Fujita T 1993 A novel mechanism for disposing of effete epithelial cells in the small intestine of guinea pigs. Gastroenterology 105:1089–1097

Johnson IT 2002 Anticarcinogenic effects of diet-related apoptosis in the colorectal mucosa. Food Chem Toxicol 40:1171–1178

Kruidenier L, Kuiper I, Lamers CB, Verspaget HW 2003 Intestinal oxidative damage in inflammatory bowel disease: semi-quantification, localization, and association with muco antioxidants. J Pathol 201:28–36

Lapre JA, De Vries HT, Termont DS, Kleibeuker JH, de Vries EG, van der Meer R 1993 Mechanism of the protective effect of supplemental dietary calcium on cytolytic activity of fecal water. Cancer Res 53:248–253

Madara JL 1990 Maintenance of the macromolecular barrier at cell extrusion sites in intestinal epithelium: physiological rearrangement of tight junctions. J Membr Biol 116:177–184

Mariadason JM, Catto-Smith A, Gibson PR 1999 Modulation of distal colonic epithelial barrier function by dietary fibre in normal rats. Gut 44:394–399

Mariadason JM, Rickard KL, Barkla DH, Augenlicht LH, Gibson PR 2000 Divergent phenotypic patterns and commitment to apoptosis of Caco-2 cells during spontaneous and butyrate-induced differentiation. J Cell Physiol 183:347–354

Mariadason JM, Velcich A, Wilson AJ, Augenlicht LH, Gibson PR 2001 Resistance to butyrate-induced cell differentiation and apoptosis during spontaneous Caco-2 cell differentiation. Gastroenterology 120:889–899

McIntyre A, Young GP, Taranto T, Gibson PR, Ward P 1993 Fermentation products of dietary fiber and protection against large bowel cancer in a rat model. Gut 34:386–391

Ohd JF, Wikstrom K, Sjolander A 2000 Leukotrienes induce cell-survival signalling in intestinal epithelial cells. Gastroenterology 119:1007–1018

Proskuryakov SY, Konoplyannikov AG, Gabai VL 2003 Necrosis: a specific form of programmed cell death? Exp Cell Res 283:1–16

Ricchi P, Zarrilli R, Di Palma A, Acquaviva AM 2003 Nonsteroidal anti-inflammatory drugs in colorectal cancer: from prevention to therapy. Br J Cancer 88:803–807

Schmitz H, Barmeyer C, Fromm M et al 1999 Altered tight junction structure contributes to the impaired epithelial barrier function in ulcerative colitis. Gastroenterology 116: 301–309

Sesink AL, Termont DS, Kleibeuker JH, Van der Meer R 1999 Red meat and colon cancer: the cytotoxic and hyperproliferative effects of dietary heme. Cancer Res 59:5704–5709

Shibahara T, Sata N, Waguri S 1995 The fate of effete epithelial cells at the villus tips of the human small intestine. Arch Histol Cytol 58:205–219

Strater J, Koretz K, Gunthert AR, Moller P 1995 In situ detection of enterocytic apoptosis in normal colonic mucosa and in familial adenomatous polyposis. Gut 37:819–825
Strater J, Wedding U, Barth TF, Koretz K, Elsing C, Moller P 1996 Rapid onset of apoptosis in vitro follows disruption of beta 1-integrin/matrix interactions in human colonic crypt cells. 110:1776–1784
Van Lieshout EMM, Van Doesburg WV, Van der Meer R 2004 Real-time PCR of host DNA in feces to study differential exfoliation of colonocytes between rats and humans. Scand J Gastroenterol, in press
Young GP, Folino M, McIntyre A et al 1996 Wheat bran suppresses potato-starch potentiated tumorigenesis at the aberrant crypt stage in a rat model of colorectal cancer. Gastroenterology 100:508–514
Yukawa M, Iizuka M, Horie Y et al 2002 Systemic and local evidence of increased Fas-mediated apoptosis in ulcerative colitis. Int J Colorectal Dis 17:70–76

DISCUSSION

Meddings: I was intrigued by the microscopic colitis. In patients with laxative abuse, where there is lots of epithelial cell death and resorption of the apoptotic cells, I have always assumed that this is why we see melanosis coli. But I have never seen melanosis coli in lymphocytic colitis, despite all the apoptosis that takes place.

Gibson: One reason is that these patients tend not to take laxatives.

Pavli: It is actually the anthraquinone that is causing the apoptosis and giving the pigmentation.

Parkos: In biopsies, I have always seen what I assume to be these apoptotic cells in the crypt. Are you saying that these are necrotic cells?

Gibson: No, the cells you are describing have histological features of apoptosis, which we have seen in the surface.

Parkos: I have seen these in the crypts, too.

Gibson: This is seen in graft-versus-host disease (GVHD).

Parkos: We see them increased in abnormal conditions. But I was under the impression that they can be seen under normal conditions. For example in a cancer resection, if you look along the crypts, you can find them periodically.

Gibson: Are they real?

Parkos: That is what I am asking.

Gibson: We were surprised, because the dogma was that apoptosis occurs in these abnormal mutated cells in the crypts, but we haven't been able to see them. We have used the Cytodeath antibody and can't see any. This was in a large number of biopsies by a highly competent and experienced electron microscopist. Of course, if you give rats dimethylhydrazine you get them all over the place, and apoptotic cells are found in the crypts in GVHD or ulcerative colitis (UC).

Parkos: In another of your pictures I thought I saw neutrophils in a crypt.

Gibson: This was in UC. There were some difficulties in this study because there was a lot of potential for artefacts. What it beautifully showed was that little focal erosions may be more important for the overall barrier defect than diffuse abnormalities in the tight junctions. We are not talking about when there are lots of neutrophils in there: we are talking about in the early stage. This might have implications if you are going to look for the cause of increased epithelial permeability that predisposes to Crohn's disease, for example. If we are looking for tight junction abnormalities we might be wrong: we might have to look for focal things that are much harder to follow.

Ghosh: You did some butyrate experiments with Caco-2 cells. Caco-2 cells are constitutively rather anti-apoptotic, as they have to be, being a cell line. How far do you think the results of Caco-2 cell experimentation are relevant to what happens on the colonic surface?

Gibson: It is just a model. We have been presenting a hypothesis and showing some evidence for it. The idea is then to find out the mechanisms and to try to find tools to demonstrate this *in vivo*. Caco-2 cells behave very much like epithelial cells in many systems, but, if you look at them histologically they do not closely resemble normal epithelial cells.

Wright: I've been aware of David Barkla's studies showing a pattern of dead and dying cells in the base of the crypt for a long time. I've always been puzzled as to what they meant. There is also the immortal strand hypothesis of Cairns where asymmetric stem cell divisions in the basal crypt occur with a mechanism for segregating any damaged DNA (Cairns 2002). I thought this was a 'pie in the sky' idea, but there was a recent paper by Chris Potten (Potten et al 2002) showing fairly clear evidence for the immortal strand hypothesis. A corollary of this is that if you then get sufficient damage to the immortal strand in the stem cells, these then die. Do you have any concept of the rather large intellectual leap between saying you have a lot of genetic defects and altruistic suicide in stem cells, or is this still a black box?

Gibson: I find it hard to believe that these are abnormal cells. Do mutations occur that frequently?

Wright: What the immortal strand hypothesis says is that numbers of mutations are built up over a period, and the stem cells eventually commit suicide. The other question concerned butyrate. Robert Goodlad in my lab showed that when you feed fermentable fibre to germ-free animals you get no effect on the hind gut. If you then expose them to bacteria you do get a trophic effect (Goodlad et al 1989). We have always assumed, from work by groups who have created skin sacs and infused butyrate at various concentrations into the colon, that it is the short chain fatty acids that have caused the proliferative effect (Kripke et al 1989). I noticed you were saying that you thought this may be due to fructose or other factors rather than short-chain fatty acids.

Gibson: That is different to the correction of atrophy in a diverted colon.

Wright: So do you think short-chain fatty acids are trophic?

Gibson: Yes, they are. We have done experiments in animal models with a chronically intubated colon. Butyrate does correct the atrophy, but the dose has to be increased with time to sustain an effect. Presumably the cells are starting to behave differently in response to butyrate. The other thing is that you can get exactly the same responses with methylcellulose or cellulose, both of which are not fermented and lead to very low butyrate concentrations. Thus there is a physical effect also. You can't ascribe everything to short-chain fatty acids.

Rhodes: Peter, you were asking how common DNA or chromosome abnormalities are in the normal epithelium. I think the answer is that they are quite common. Terri Brentnall and her group produced some nice studies using fluorescent *in situ* hybridization for large chunks of DNA, including p53 and the centromere of chromosome 17. They showed major deletions or duplications occurring in about 10% of the surface epithelial cells in the normal colon, rising to 20–35% in the dysplastic colitic colon. How much apoptosis do you see in the colitic colon? One of the major mechanisms for cancer development in IBD could be failure of apoptosis. This might be relative rather than absolute failure. Is there an increase or reduction in net numbers of apoptotic cells?

Gibson: There is an increase in active disease. It is hard to know what the quantification means.

Rhodes: Terri Brentnall showed roughly a threefold increase in major chromosomal abnormalities. Is there concomitant threefold increase in apoptosis?

Gibson: You can't do this comparison. But once the cells are up on the surface they are not going to cause trouble are they?

Rhodes: I don't know.

Wright: If you believe Vogelsteins's data, there is a top down hypothesis we could debate (Shih et al 2001).

Meddings: Isn't the fact that you see those cells with chromosomal abnormalities reflecting a relative failure of apoptosis?

Rhodes: Yes, it is. It is not easy to fit this with an apparent increase.

Gibson: We see apoptotic cells in the normal small intestinal crypt, but not in the colonic crypt.

Wright: It is less likely to be a failure of apoptosis; more a successful stochastic expansion of a mutated stem cell line, with a growth advantage.

Rhodes: How do you then tie in the cancer risk with the duration and extent of inflammation?

Wright: You make a general mutation rate as a proportion of the cell proliferation rate. Crypt fission has increased greatly in UC.

Rhodes: I thought you weren't too keen on the general principle of increased proliferation as a basic starting point for cancer.

Wright: I'm ambivalent about it now!

Pavli: I want to ask about the focal abnormalities in epithelial cell barrier function. There is quite an old model looking at rabbits using anthraquinone derivatives. This has been mainly used to look at macrophage phagocytosis of the apoptotic epithelial cells. Has anyone used that model to look at barrier function in epithelial cells?

Gibson: No. In that model the macrophages are in the epithelium itself. This is the only animal I know of where macrophages are in the epithelial layer.

Pavli: I think there are other models where macrophages can grab the apoptotic cells through the basement membrane and phagocytose them.

Meddings: There are some publications looking at induction of apoptosis (Sun et al 1998) and showing overall increased permeability. This is also true in some infective conditions such as *Giardia*: if apoptosis is inhibited the increased permeability effect of infection is alleviated (Chin et al 2002).

Gibson: I would have thought that one cell alone could not do it. You would need a nest. I think that is what this study showed. Whether that is what is actually happening in UC I cannot be sure.

Powrie: What about the idea that a basal level of apoptosis might be important for some mechanisms of intestinal homeostasis? Some data from Gordon Macpherson are relevant to this. He showed that intestinal dendritic cells (DCs) can take up apoptotic epithelial cells which are able then to travel to the mesenteric lymph nodes. He has postulated that these have more tolerogenic properties. Of course, apoptosis is associated in some systems with IL10 and TGFβ production, which we know are important in intestinal homeostasis.

Gibson: The big problem with this is in the human, how did the apoptotic cells get to the macrophages?

Powrie: The DCs could go right up into the epithelium.

Gibson: We can't see them.

Powrie: We see them in the mouse. What is the consensus in humans? Are intra-epithelial DCs found in the colon?

Furrie: I've looked and can't find them. I have looked at quite a lot of biopsies.

Gibson: We can't see them. In fact, they don't need to go there; they can pop through the fenestrae.

Uhlig: These cells are associated with lymphoid clusters but are rare or absent in the non-follicle associated epithelium.

Gibson: We didn't look near follicles.

Ghosh: We have kept on revisiting this story about the sterility of the crypts. We heard yesterday about the NOD2 expression in Paneth cells. As intracellular sensors for muramyl dipeptide, the base of the crypt doesn't seem the right place for this to be situated in if it is relatively sterile. Caco-2 cells do express NOD2. To what extent is this expression of NOD2 in Paneth cells a quantitative phenomenon rather than a purely qualitative all or none?

Mahida: One issue we need to consider is that in the normal colonic crypt there will usually be no Paneth cells. It has been suggested that one of the functions of the antimicrobial peptides derived from the Paneth cells may be to protect stem cells. As regards the role of NOD2 in Paneth cells, I think this is an open question. There may be bacterial products in the lumen in the absence of bacteria. There is also some old literature from the 1970s showing that Paneth cells can actively phagocytose microorganisms (Erlandsen & Chase 1972). As far as I am aware, this hasn't been looked at recently.

Sartor: Just because there are no bacteria there doesn't mean that there are no bacterial products. Bacteria continuously secrete peptidoglycan that has to be cross-linked in order to become part of the cell wall. This could get into the crypt.

Wright: I have a related question concerning the idea of the resistance or not, or sensitivity of crypt basal stem cells to death processes. Chris Potten has shown that the sensitivity to radiation of the stem cells in the base of the crypt is far lower than transit amplifying cells. 2 μCi/g of bodyweight can induce apoptosis in these stem cells: you said that you don't see phagocytosis of the debris from these cells by epithelial cells but in 1974 this was shown after causing cell death by 2 μCi/g bodyweight of tritiated thymidine. Adjacent cells phagocytosed that labelled debris and then differentiated into endocrine, Paneth, goblet and enterocytes (Cheng & Leblond 1974).

Gibson: We can't see any evidence in the normal colon. There was a report in patients with familial adenomatous polyposis that apoptotic cells were being phagocytosed by adjacent epithelial cells, but, in that study, phagocytosis was not seen in the crypt, despite the abnormal proliferation and other events. Epithelial cells are probably capable of doing it, it is just that in a normal colon it doesn't seem to be relevant.

Sartor: Before we hear Ingvar Bjarnason's presentation about whether there is a genetic basis for IBD, in your human UC experiments do you think that the apoptosis/necrosis that you are seeing is a consequence of inflammation, or a primary event? Did you see any evidence of heterogeneity in your patients that would suggest that some patients arrive there by a primary epithelial defect, while others have an immunoregulatory defect?

Gibson: If you can tell how to distinguish between the chicken and the egg, I'll give you the answer. I think the inflammation is a secondary event.

References

Cairns J 2002 Somatic stem cells and the kinetics of mutagenesis and carcinogenesis. Proc Natl Acad Sci USA 99:10567–10570

Cheng H, Leblond CP 1974 Origin, differentiation and renewal of the four main epithelial cell types in the mouse small intestine. V. Unitarian Theory of the origin of the four epithelial cell types. Am J Anat 141:537–561

Chin AC, Teoh DA, Scott KG, Meddings JB, Macnaughton WK, Buret AG 2002 Strain-dependent induction of enterocyte apoptosis by Giardia lamblia disrupts epithelial barrier function in a caspase-3-dependent manner. Infect Immun 70:3673–3680

Erlandsen SL, Chase DG 1972 Paneth cell function: phagocytosis and intracellular digestion of intestinal microorganisms. II. Spiral microorganisms. Ultrastruct Res 41:319–333

Goodlad RA, Ratcliffe B, Fordham JP, Wright NA 1989 Does dietary fibre stimulate intestinal epithelial cell proliferation in germ free rats? Gut 30:820–825

Kripke SA, Fox AD, Berman JM, Settle RG, Rombeau JL 1989 Stimulation of intestinal mucosal growth with intracolonic infusion of short-chain fatty acids. J Parenter Enteral Nutr 13:109–116

Potten CS, Owen G, Booth D 2002 Intestinal stem cells protect their genome by selective segregation of template DNA strands. J Cell Sci 115:2381–2388

Shih IM, Wang TL, Traverso G et al 2001 Top-down morphogenesis of colorectal tumors. Proc Natl Acad Sci USA 98:2640–2645

Sun Z, Wang X, Wallen R et al 1998 The influence of apoptosis on intestinal barrier integrity in rats. Scand J Gastroenterol 33:415–422

Genetic aspects of intestinal permeability in inflammatory bowel disease

Ken Takeuchi, Laurence Maiden and Ingvar Bjarnason[1]

Department of Medicine, Guy's, King's, St Thomas' School of Medicine, Bessemer Road, London SE5 9PJ, UK

Abstract. There is a long-standing belief that disruption of the intestinal barrier function may lead to systemic and local intestinal disease. The role of increased intestinal permeability in Crohn's disease is reviewed here. What is not in doubt is that intestinal permeability in patients with Crohn's disease is increased proportional to disease activity; it can be used to predict clinical relapse of disease and prognosis; and a small proportion of first-degree relatives have increased intestinal permeability. This last finding has been subject to much speculation. In particular it has been suggested that it represents a genetically determined abnormality. If so it might play an important pathogenic process in the disease. However this permeability change in relatives does not conform to a classical inheritance pattern and in some studies it is found in the patients' spouses. This suggests an environmental cause for the changes. However proponents of an environmental factor have been singularly inactive in attempting to identify this agent(s). In view of recent research it seems likely that the increased intestinal permeability in relatives of Crohn's patients may be secondary to sub-clinical intestinal inflammation. This inflammation conforms to an inherited additive trait. The genetic basis for this inflammation is being studied.

2004 Inflammatory bowel disease — crossroads of microbes, epithelium and immune systems. Wiley, Chichester (Novartis Foundation Symposium 263) p 151–163

One of the more simplistic, but useful, models for understanding the pathogenesis of intestinal diseases compartmentalizes the gut into (i) lumenal aggressors, (ii) intestinal barrier function and (iii) mucosal defence. The lumenal aggressors are responsible for the exceptional predicament in which the intestinal tract finds itself, whereby it is exposed to a more inhospitable milieu than any other organ of the body. The intestine (small and large) is exposed to a variety of ingested

[1]This paper was presented at the symposium by Ingvar Bjarnason to whom correspondence should be addressed.

foodstuffs, digestive enzymes and bile in addition to bacteria and their degradation products. The gastrointestinal mucosa has adapted to the aggressors by restricting their access to the surface epithelium, maintaining an effective barrier to their permeation and by involvement of local and systemic, innate and induced, biochemical and immunological responses when the aggressors gain access to the mucosa. Although there is no one single factor more important than another in the maintenance of normal intestinal function and health, there is a prevailing and long-standing suggestion that lumenal aggressive factors and intestinal integrity is important for many local and systemic diseases. Nowhere is this better highlighted that in patients with classical inflammatory bowel disease (IBD). Almost every conceivable luminal aggressive factor has been implicated in its cause. Furthermore, even before it was possible to measure intestinal integrity (permeability) in humans non-invasively and reliably it was suggested that increased intestinal permeability was the culprit to the development of intestinal (Shorter et al 1972) and systemic diseases (Hong & Amman 1972, Svartz 1972, Bywaters 1988). When intestinal permeability was found to be increased in patients with Crohn's disease (Ukabam et al 1982, Bjarnason et al 1983) and colonic permeability in ulcerative colitis (Rask-Madsen & Schwartz 1979) almost all aspects of this disease were ascribed to the loss of the intestinal barrier.

The intestinal defence mechanisms act in a highly coordinated synergistic fashion in order to limit the effects of luminal aggressive factors on the mucosa. Interestingly, it has been pointed out that there is a striking conformity in measured permeability change and inflammation in a number of disparate diseases of the small bowel (Bjarnason et al 2004). Accordingly (Table 1), disorders characterized as 'barrier breakers', infections and immune deficiency are all associated with a two–fivefold increase in permeability and inflammation.

This suggests that by whatever means by which the gut is damaged the intestinal permeability–inflammation reaction (either can be the cause of the other) is predetermined and overshadows the changes caused by the initial assault. However when the permeability and inflammatory changes are analysed in patients with Crohn's disease some interesting departures from this picture are evident. Here we will review much of the data on intestinal permeability in Crohn's disease and assess suggestions that it may represent a genetically determined abnormality that plays an important pathogenic role in the disease.

Intestinal permeability in Crohn's disease

What is not in doubt is that intestinal permeability is increased in patients with small bowel involvement of Crohn's disease and in half of those with Crohn's colitis without apparent small bowel involvement (Ukabam et al 1982, Bjarnason

TABLE 1 Association between intestinal permeability and inflammation

Condition	*Permeability increase*	*Inflammatory increase*
Barrier breakers		
NSAIDs	2–5	3–5
Alcohol	2–5	3
Renal failure	2–3	5
Radiation	2–3	3
Cytotoxic drugs	2–4	NA
Ischaemia–exercise	2	NA
Endotoxins	2–4	NA
Lumenal aggressors		
Salmonella–Shigella	2–5	8
Ascaris lunbricoides	2–3	NA
Rotavirus	2–8	NA
'Gastroenteritis'	2–5	3
Tropical enteropathy Travellers diarrhoea	2–5	NA
Mucosal defence		
HIV-AIDS	2–10	2–4
Hypogammaglobulinaemia	3	7
Pouchitis	2	2–5
Miscellaneous		
Cystic fibrosis	3–5	NA
Diabetic diarrhoea	2–3	2
Coeliac disease	2–5	2
Crohn's disease		
Inactive	2–3	2–4
Active	3–5	15–20

et al 1983, Casellas et al 1986, Murphy et al 1989, Pironi et al 1990, Adenis et al 1992, Bjarnason et al 1995). The permeability increase is similar to that found in other enteropathies (about two–threefold increase) (Bjarnason et al 2004). This has led to the suggestion that tests of intestinal permeability might serve as screening tests to distinguish between out-patients with irritable bowel syndrome and IBD that present to gastroenterologists (Tibble et al 2002). As it turns out the differential urinary excretion of lactulose/L-rhamnose has about 90% sensitivity in detecting active Crohn's disease of the small bowel, but the specificity is very poor, a whole variety of small bowel diseases demonstrating, somewhat predictably (Bjarnason et al 1995), increased intestinal permeability (Tibble et al 2002).

The increased intestinal permeability in Crohn's disease relates to the site and extent of disease and clinical disease activity (Pironi et al 1990, Teahon et al 1991, 1993). The tests are good predictors of clinical disease activity if studied sequentially, but because of the differences in disease location and extent of disease between patients a single test result can not give reliable information of clinical disease activity.

One of the major strengths of intestinal permeability testing in Crohn's disease is the universal agreement that a normal test result in asymptomatic patients heralds a good short-term (6–12 months) prognosis while increased intestinal permeability is predictive of a clinical relapse. The first study assessed patients with small bowel Crohn's disease that had been successfully treated with an elemental diet for 4–6 weeks. Those who had increased intestinal permeability at the end of the treatment had a clinical relapse of their disease within 6 months (Teahon et al 1993). Those with normal intestinal permeability, however, faired well and maintained their remission. Wyatt et al (1993) studied a more random group of patients with Crohn's disease (small and large bowel) and found that those prone to clinical relapse had a significantly greater increase in small intestinal permeability than those maintaining clinical remission. An additional four studies (D'Inca et al 1999, Hilsden et al 1999, Arnott et al 2000, Jorgensen et al 2001) showed the same, further confirming the value of permeability estimations to predict clinical relapse of disease.

Intestinal permeability in first-degree relatives

There is a long-standing belief that increased intestinal permeability may play an important pathogenic or even a causative role in small bowel Crohn's disease. It was therefore only a matter of time until first-degree relatives of patients with Crohn's disease were studied. The first study was that of Hollander et al (1986). They used polyethylene glycol, mean molecular weight 400 Da (PEG 400). Not withstanding all the criticism that has been aimed at the PEG 400 method these workers found that the vast majority of patients and their first-degree relatives had increased urinary excretion of PEG 400 as compared with the control group. Taken at face value this opened up a new research field and suggested that increased intestinal permeability might be an underlying genetically determined pathogenic process in Crohn's disease. Such a sub-clinical disease marker would define a genotype which does not progress to the full disease (phenotype) because other genetic or environmental factors were lacking. However two subsequent studies (Ruttenberg et al 1992, Teahon et al 1992) showed normal PEG 400 permeability in

TABLE 2 Intestinal permeability studies in first-degree relatives of patients with Crohn's disease

Authors	*Number of patients/ relatives studied*	*Number of patients/ relatives abnormal*
Ainsworth et al 1989	15/20	2 (13%)/not reported
Teahon et al 1992	28/32	11 (39%)/not reported
May et al 1993	36/38	13 (36%)/6 (16%)
Katz et al 1989	25/41	Not reported
Yacyshyn & Meddings 1995	15/13	10 (67%)/7 (54%)
Secondulfo et al 2001	16/26	6 (37.5%)/3 (11.5%)
Howden et al 1994		Normal in all
Peeters et al 1997	25/67	13 (52%)/17 (25%)
Soderholm et al 1999	39/34	14 (36%)/6 (18%)

Crohn's disease relatives and indeed reduced, not increased, PEG 400 permeability in the patients. Furthermore no significant differences in PEG 400 permeation were found between the groups of patients, relatives and controls, or between diseased and healthy relatives (Munkholm et al 1994). However, in the meantime a number of studies had addressed the possibility that the relatives might have increased intestinal permeability as assessed by the differential urinary excretion of di-/monosaccharides or [^{51}Cr]-EDTA. Table 2 shows some of these results.

With minor discrepancy it can be concluded that perhaps 10–25% of the relatives had increased intestinal permeability although one study showed a prevalence of just over 50% (Yacyshyn & Meddings 1995). This study also suggested that relatives who had increased permeability were found to have increased CD45RO expression of peripheral B cells. This was taken as evidence of a systemic reaction to the luminal permeation of antigens consequent to the permeability changes. Others have not found convincing correlations between increased intestinal permeability in the relatives and anti-*Saccharomyces* antibodies (Secondulfo et al 2001, Vermiere et al 2001). Although there is one case report that shows a woman with a positive family history of Crohn's disease who had been confirmed with increased intestinal permeability to [^{51}Cr]-EDTA eight years before being diagnosed with Crohn's disease (Irvine & Marshall 2000), the vast majority of the 'abnormal' family members do not develop Crohn's disease.

The main limitation of the family studies is that they only examined a very small number of available relatives. A Belgian study examined whole families of Crohn's disease patients. Here it was suggested that they found a clustering of cases within certain families and normal intestinal permeability in others (Peeters et al 1997).

However, analyses of affected cases did not disclose any particular inheritance pattern. In a similar study from Iceland assessing whole families, in which over 60% of the first-degree relatives were studied, it was shown that only 18% had increased intestinal permeability and again there was no hint of a particular type of inheritance.

Another finding of interest was that administration of aspirin (Hilsden et al 1996) or nonsteroidal anti-inflammatory drugs (NSAIDs) (Zamora et al 1999) to the first-degree relatives was followed by a greater permeability increase than when given to volunteers.

An apparent drawback to the interpretation of these studies was suggested by the findings of increased intestinal permeability in apparently healthy spouses of patients with Crohn's disease. Three studies show 23% (Soderholm et al 1999), 13% (Breslin et al 2001) and 32% (Peeters et al 1997) of the spouses to have increased intestinal permeability. This suggested that abnormal permeability is likely to reflect shared exposure to environmental factors. However, in the Icelandic study the spouses were normal. It would seem that the 'environmental' investigators have the edge on those that believe in the genetic aspects at present. It is however a matter of considerable curiosity that the former have made no attempt whatsoever to identify what these factors might be.

Intestinal inflammation in first-degree relatives

Increased intestinal permeability can be the cause or consequence of intestinal inflammation (Bjarnason et al 1995). Being firm believers that increased intestinal permeability is not a primary, genetically-determined abnormality in Crohn's disease, we assessed the possibility that first-degree relatives might have subclinical intestinal inflammation. The study was carried out in Iceland which is ideally suited for genetic research because the origins of the population are well defined and they lived in considerable isolation until the mid 20th century. The study looked at a quarter of all patients with Crohn's disease in the country (Thjodleifsson et al 2003). Complete families of these patients were studied in relation to intestinal permeability, various serological antibodies and the presence of intestinal inflammation. Intestinal inflammation was assessed with a well-validated marker, namely faecal calprotectin (Sigthorsson et al 1998, Tibble et al 1999, 2000). Approximately 50% of first-degree relatives had evidence of subclinical intestinal inflammation. More importantly the inheritance pattern conformed to an additive trait and most spouses were normal. The authors suggested a number of possible genetic mechanisms that might lead to such inflammation. It seems likely from this study that the permeability abnormalities that are described in first-degree relatives are the consequence rather than the cause of this inflammation.

References

Adenis A, Colombel JF, Lecouffe P et al 1992 Increased pulmonary and intestinal permeability in Crohn's disease. Gut 33:678–682

Ainsworth M, Eriksen J, Rasmussen JW, Schaffalitzkydemuckadel OB 1989 Intestinal permeability of 51Cr-labelled ethylenediaminetetraacetic acid in patients with Crohn's disease and their first degree relatives. Scand J Gastroenterol 24:993–998

Arnott ID, Kingstone K, Ghosh S 2000 Abnormal intestinal permeability predicts relapse in inactive Crohn disease. Scand J Gastroenterol 35:1163–1160

Bjarnason I, O'Morain C, Levi AJ, Peters TJ 1983 The absorption of 51Cr EDTA in inflammatory bowel disease. Gastroenterology 85:318–322

Bjarnason I, Macpherson AJM, Hollander D 1995 Intestinal permeability: an overview. Gastroenterology 108:1566–1581

Bjarnason I, Takeuchi K, Bjarnason A, Adler SN, Teahon K 2004 The G.U.T. of gut. Scand J Gastroenterol, in press

Breslin NP, Nash C, Hilsden RJ et al 2001 Intestinal permeability is increased in a proportion of spouses of patients with Crohn's disease. Am J Gastroenterol 96:2934–2938

Bywaters EGL 1988 Historical aspects of the etiology of rheumatoid arthritis. Br J Rheumatol 27(suppl 2):110–115

Casellas F, Aguade S, Soriano B, Accarino A, Molero J, Guarner L 1986 Intestinal permeability to 99mTc diethylene-tetraaminopentaacetic acid in inflammatory bowel disease. Am J Gastroenterol 81:767–770

D'Inca R, Di Leo V, Corrao G et al 1999 Intestinal permeability test as a predictor of clinical course in Crohn's disease. Am J Gastroenterol 94:2956–2960

Hilsden RJ, Meddings JB, Sutherland LR 1996 Intestinal permeability changes in response to acetylsalicylic acid in relatives of patients with Crohn's disease. Gastroenterology 110:1395–1403

Hilsden RJ, Meddings JB, Hardin J, Gall DG, Sutherland LR 1999 Intestinal permeability and postheparin plasma diamine oxidase activity in the prediction of Crohn's disease relapse. Inflamm Bowel Dis 5:85–91

Hollander D, Vadheim C, Brettholz E, Pattersen GM, Delahunty T, Rotter JI 1986 Increased intestinal permeability in patients with Crohn's disease and their relatives. Ann Int Med 105:883–885

Hong R, Amman AJ 1972 Selective absence of IgA: autoimmune phenomena and autoimmune diseases. Human Pathol 69:451–496

Howden CW, Gillanders I, Morris AJ, Duncan A, Danesh B, Russell RI 1994 Intestinal permeability in patients with Crohn's disease and their first-degree relatives. Am J Gastroenterol 89:1175–1176

Irvine RJ, Marshall JK 2000 Increased intestinal permeability preceeds the onset of Crohn's disease in a subject with familial risk. Gastroenterology 119:1740–1744

Jorgensen J, Ranlov PJ, Bjerrum PJ, Diemer H, Bisgaard K, Elsborg L 2001 Is an increased intestinal permeability a valid predictor of relapse in Crohn disease? Scand J Gastroenterol 36:521–527

Katz KD, Hollander D, Vadheim CM et al 1989 Intestinal permeability in patients with Crohn's disease and their healthy relatives. Gastroenterology 97:927–931

May GR, Sutherland LR, Meddings JB 1993 Is small intestinal permeability really increased in relatives of patients with Crohn's disease? Gastroenterology 104:1627–1632

Munkholm P, Langholz E, Hollander D et al 1994 Intestinal permeability in patients with Crohn's disease and ulcerative colitis and their first degree relatives. Gut 35:68–72

Murphy MS, Eastham EJ, Nelson R, Pearson ADJ, Laker MF 1989 Intestinal permeability in Crohn's disease. Arch Dis Child 64:321–325

Peeters M, Geypens B, Claus D et al 1997 Clustering of increased intestinal permeability in families with Crohn's disease. Gastroenterology 113:802–807

Pironi L, Miglioli M, Ruggeri E et al 1990 Relationship between intestinal permeability to (51Cr)EDTA and inflammatory activity in asymptomatic patients with Crohn's disease. Dig Dis Sci 35:582–588

Rask-Madsen J, Schwartz M 1979 Absorption of 51CrEDTA in ulcerative colitis following rectal instillation. Scand J Gastroenterol 5:361–368

Ruttenberg D, Young GO, Wright JP, Isaacs S 1992 PEG 400 excretion in patients with Crohn's disease, their first degree relatives, and healthy volunteers. Dig Dis Sci 37: 705–708

Secondulfo M, de Magistris L, Fiandra R et al 2001 Intestinal permeability in Crohn's disease patients and their first degree relatives. Dig Liver Dis 33:649–651

Shorter RG, Huizenga GA, Spencer RJ 1972 A working hypothesis for the etiology and pathogenesis of nonspecific inflammatory bowel disease. Dig Dis Sci 17:1024–1031

Sigthorsson G, Tibble J, Hayllar J et al 1998 Intestinal permeability and inflammation in patients on NSAIDs. Gut 43:506–511

Soderholm JD, Olaison G, Lindberg E et al 1999 Different intestinal permeability patterns in relatives and spouses of patients with Crohn's disease: an inherited direct in mucosal defence? Gut 44:96–100

Svartz N 1972 The primary cause of rheumatoid arthritis is an infection. The infectious agent exists in milk. Acta Med Scand 192:231–239

Teahon K, Smethurst P, Levi AJ, Bjarnason I 1991 The effect of elemental diet on intestinal permeability and inflammation in Crohn's disease. Gastroenterology 101: 84–89

Teahon K, Smethurst P, Levi AJ, Menzies IS, Bjarnason I 1992 Intestinal permeability in patients with Crohn's disease and their first degree relatives. Gut 33:320–323

Teahon K, Smethurst P, Macpherson AJ, Levi AJ, Menzies IS, Bjarnason I 1993 Intestinal permeability in Crohn's disease and its relation to disease activity and relapse following treatment with elemental diet. Eur J Gastroenterol Hepatol 5: 79–84

Thjodleifsson B, Sigthorsson G, Cariglia N et al 2003 Subclinical intestinal inflammation: an inherited abnormality in Crohn's disease relatives? Gastroenterology 124:1728–1737

Tibble J, Sigthorsson G, Foster R et al 1999 High prevalence of NSAID enteropathy as shown by a simple faecal test. Gut 45:362–366

Tibble J, Teahon K, Thjodleifsson B et al 2000 A simple method for assessing intestinal inflammation. Gut 47:506–513

Tibble J, Sigthorsson G, Foster R, Forgacs I, Bjarnason I 2002 Use of surrogate markers of inflammation and Rome criteria to distinguish organic from nonorganic intestinal disease. Gastroenterology 123:450–460

Ukabam SO, Clamp JR, Cooper BT 1982 Abnormal intestinal permeability to sugars in patients with Crohn's disease of the terminal ileum and colon. Digestion 27:70–74

Vermiere S, Peeters M, Vlietinck R et al 2001 Anti-Saccharomyces antibodies (ASCA), phenotypes of IBD, and intestinal permeability: a study in IBD families. Inflamm Bowel Dis 7:8–15

Wyatt J, Vogelsang H, Hubl W, Waldhoer T, Lochs H 1993 Intestinal permeability and the predictor of relapse in Crohn's disease. Lancet 341:1437–1439

Yacyshyn BR, Meddings JB 1995 CD45RO expression on circulating CD19+B cells in Crohn's disease correlates with intestinal permeability. Gastroenterology 108:132–137

Zamora SA, Hilsden RJ, Meddings JB, Butzner JD, Scott RB, Sutherland LR 1999 Intestinal permeability before and after ibuprofen in families of children with Crohn's disease. Can J Gastroenterol 13:31–36

DISCUSSION

Parkos: Have you looked at any biopsies from asymptomatic patients, to see whether you can see neutrophils in the epithelium, for example?

Bjarnason: We have complete access to these patients, but we haven't asked them for any biopsies yet. They have undergone some localization studies. Half had increased faecal calprotectin levels and half had normal values. They underwent a white cell scan. We found that half of the patients with increased faecal calprotectin have diffuse small bowel inflammation. The next stage is for them to have a capsule enteroscopy so we can see the inflammation.

Parkos: So the key question is: what keeps everyone from having massive neutrophil migration into their bowel? A large percentage of circulating neutrophils end up there.

Bjarnason: I am not sure. If you look at the white cell excretion, only about 1% come out in the stool in normals. Why don't they go there? I think this is down to the permeability–inflammation interaction in so-called normals. I think if we were germ-free we would have zero inflammation, but in the state that we are all in, which I refer to as low-grade IBD, we have about 1% going to the gut. It doesn't matter how you destroy the gut barrier — radiation, NSAIDs or alcohol, for example — the response after you allow the luminal contents into the mucosa is predetermined. You send in two-to-four times more neutrophils. The common final pathway in all these enteropathies is an interaction between gut flora and the mucosal defence system.

Jewell: John Meddings, I believe in your cohort of first degree relatives that had the abnormal permeability, there was a plan to follow them up to see whether they subsequently developed Crohn's disease. Do you have an answer that you can share with us?

Meddings: I can't. We have gone back and looked, and unfortunately Calgary is an incredibly mobile population, and in the 10 years since the first study we could only find a few scattered people. That group has gone. There is one reported case from McMaster (from Jan Irvine and John Marshall) of a young girl who underwent a permeability study back in the 1980s (Irvine & Marshall 2000). The only risk factor for this girl was her brother who had Crohn's disease. They completely investigated that girl at the time with a colonoscopy and a small bowel biopsy. She was entirely normal. 10 years later she has classical Crohn's. This is the only reported case I am aware of where this has happened.

Bjarnason: We have one from an ankylosing spondilitis study in Iceland. We followed these patients up for four years and there is otherwise no conversion.

Meddings: I agree totally: this finding that relatives have abnormal permeability is reproducible. Spouses also have it. I cannot believe that this is a genetic abnormality. My bias is that this is an interaction between some microflora and some genetically programmed epithelial cell and/or immune system. I suspect it is early inflammation. We have just started a study to look at these patients with capsule endoscopy.

Rhodes: Can I give you a crazy hypothesis for an environmental factor? Some years ago we had a temporary nanny looking after our children in the summer holidays. We were conscious that they started to have heads of froth on their drinks. It transpired that our nanny had been using lots of detergent in the washing up and then leaving things to dry. I asked some of our patients about whether they rinse their washing up. One-third make no attempt to rinse. There are remarkably few studies on the effects of low-level detergent. One study done by Charles Clark showed a significant increase in intestinal permeability and some changes in histology in rodents fed what he claimed was a level of detergent that a baby might be exposed to (Mercurius-Taylor e al 1984). There was a study from the 1940s showing that rabbits fed detergent developed atheroma within six weeks (Hueper 1944). There have hardly been any studies done. If you look at enteral feeding studies in Crohn's disease, there are strange discrepancies between the therapeutic responses to different polymeric feeds. We had thought that this might be due to the fat content, but the evidence for this is weak. Some of these feeds usually have an emulsifier in them, particularly the ones that have a reasonable amount of fat and come ready mixed in the can. This is usually a detergent. There are similar detergents in margarine and processed foods. It all fits epidemiologically.

Jewell: I remember Charles Clarke floating a hypothesis to the British Society of Gastroenterology that detergents were the cause of duodenal ulcer because they disrupted the epithelium. We haven't heard much of this hypothesis.

Sartor: Why is IBD mostly distal if highest concentrations will be proximal?

Rhodes: Perhaps the most sensitive bit of the gut is the distal ileum where it is ultrapermeable to bacteria and so on.

Schölmerich: One argument against this is that we have natural detergents at a high concentration in our gut every day. Bile acids are potent detergents.

Rhodes: That wouldn't explain Charles Clarke's results.

Bjarnason: The permeability guys are ahead of you. In 1984 we were self-administering cetrimide, which is a powerful detergent. This will give a 20-fold increase in permeability. There is no clinical situation associated with higher values.

Ghosh: How well correlated was the faecal calprotectin and the raised permeability in the relatives in your study?

Bjarnason: The correlation was disappointing.

Meddings: Can I add that the permeability was a small bowel permeability measurement, whereas calprotectin could come from anywhere.

Ghosh: Was there any relationship to smoking?

Bjarnason: Smoking in these relatives was a minor issue. It was so small that we didn't bother going into the detail of it.

Ghosh: How confident are you of your assertion that alcoholic binges lead to IBD relapse?

Bjarnason: It wouldn't happen with a British drinker. We are talking about weird people like the northern races. Any of you who has a collaborator in Finland or Iceland know that we sit down with a bottle of whisky and then two hours later you ask your relatives to go and get another one. This binge drinking is well recognized as a cause of relapse for IBD in Nordic countries. There is no study to show this, although I believe it.

Jewell: You are at a disadvantage genetically in that you only had 48 patients with Crohn's disease. I know that *NOD2* mutations are much lower in Scandinavia than they are in the UK, France and Germany, for example, but have you *NOD2*-typed your patients?

Bjarnason: We have *NOD2*-typed all the IBD material in Iceland, and out of 187 Crohn's patients there were three positive. Out of that material there were none positive in the study of the 48 patients that we did our genetic study on.

Jewell: Are the three positives a single mutation?

Bjarnason: Yes, I think so.

Jewell: What are the results of genomewide screens?

Bjarnason: If you take the Crohn's disease patients in Iceland and the positive faecal calprotectin relatives, then you get a good LOD score on chromosome 16 which is different from all the others that have been described. This same hit is found also in ankylosing spondilitis (AS). Then you get into a very interesting interaction between the ileitis of AS and Crohn's disease and whether these fall within a spectrum of the same disease process. We have taken all the AS patients and Crohn's disease patients in Iceland and looked at whether they are significantly more related than any other two groups of people in Iceland. There seems indeed to be a close family link between Crohn's disease and AS. There are various observations that suggest that the ileitis of AS has some pathological and pathophysiological features that resemble Crohn's disease.

Schölmerich: Regarding the alcohol problem, there are interesting data from liver research by Christian Bode and colleagues. They have shown that people drinking high levels of alcohol have more bacterial colonization in the upper gastrointestinal tract or upper small intestine. They have bacterial translocation even in the absence of liver cirrhosis or liver disease. This can be easily measured by looking at C-reactive protein 48 h after heavy drinking. This indirect readout of bacterial translocation should be relatively easy to measure.

Roediger: Have you had the chance to look at labelled white cell scanning in the relatives that have a high permeability? Have you got other evidence of white cell migration into the lumen?

Bjarnason: We did white cell scan the relatives, but we selected those with increased inflammation, not increased permeability. Half of them have abnormal white cell scans.

Roediger: NSAIDs affect the metabolism of the small and large bowel mucosa, and we can show this with metabolic scanning (PET scanning). Have you excluded every non-steroidal from the relatives in which you found positive permeability?

Bjarnason: We were rigorous in excluding every factor which could increase permeability. Many people were excluded from the study on that basis.

Parkos: I want to make a comment with regard to neutrophil migration and permeability. We all think simplistically that when neutrophil migration occurs across the epithelium, it results in holes and causes changes in permeability. Now, there are a few reports looking at low density neutrophil migration. It is possible to get migration of leukocytes across monolayers *in vitro* without large changes in permeability. The way I think of this is that at a low level of migration the epithelium is able to have coordinated regulation, opening the junctions as neutrophils go through and then sealing them. At some point this becomes disrupted, resulting in physical discontinuities. This may explain why you can sometimes measure faecal calprotectin and not see a permeability change. It has to do with whatever that critical level of migration is. This is more difficult to regulate in our epithelial model.

Bjarnason: Are you saying that they are going through but are not activated, and therefore they don't cause damage?

Parkos: No, I think that as there are more neutrophils migrating across the monolayer, at some point the traffic becomes too robust and the epithelium can't reseal.

Bjarnason: When we quantitate the calprotectin in the relatives, some of the levels are on par with active Crohn's disease.

Parkos: I guess you could argue about whether that is happening over a large surface area or a small one. This could make a difference.

Gibson: Could I take up the challenge of identifying the environmental factors causing IBD? There are Dutch data showing that feeding rats fructo-oligosaccharides increases intestinal permeability (Bovee-Oudenhoven et al 2003). If that is occurring in humans, this is potentially a significant factor. The intake of fructose and highly fermentable polyols such as sorbitol is increasing in our community. Our hypothesis is that if dietary non-absorbed carbohydrates are changing permeability, they may be inducing inflammation. This might be the Western environmental factor you mentioned.

Bjarnason: If this was the case we wouldn't see this clustering within specific families because of the global consumption of these substances.

Gibson: Everyone differs in their response to these things.

Bjarnason: In this study we had 150 controls, and none of them showed the permeability or the calprotectin results that the relatives did. Presumably, some of the 150 controls must be eating these substances.

Gibson: We all eat them. It is a matter of how much you eat and the individual's response to them. It is not going to be the only factor: detergent consumption, or smoking status, or neonatal exposure might also have an effect.

Meddings: It is impossible to control for this, although it's a great idea. Our control group comes from the same environment as these families, and we don't see it in them. The rationale for having a population control group is that these uncontrolled variables will be accounted for as both the control and test group are exposed to them.

Gibson: We know that increased consumption of refined sugars is the pre-Crohn's disease dietary abnormality in multiple studies. This gives us a clue, although we don't know what their intake of fructose and fructans is. The key is whether the intake of these things changes intestinal permeability in the population.

Furrie: As far as ulcerative colitis (UC) and defensin expression are concerned, the studies that first showed up-regulation of inducible defensins in UC used a control group that were chronic alcoholics with alcohol-induced colitis. They don't have any up-regulation of inducible defensins. You are talking about Crohn's disease and not UC, but this will also increase permeability but doesn't show the same things as UC at a molecular level.

Jewell: What does alcoholic colitis involve?

Furrie: This was Deborah O'Neil's cohort of patients who were chronic alcoholics who had colitis. It was felt that they didn't have the features of IBD so they were used as a control group.

References

Bovee-Oudenhoven IM, ten Bruggencate SJ, Lettink-Wissink ML, van der Meer R 2003 Dietary fructo-oligosaccharides and lactulose inhibit intestinal colonisation but stimulate translocation of salmonella in rats. Gut 52:1572–1578

Hueper WC 1944 Experimental studies on the therapy and the prevention of degenerative vascular disease. II. The effects of several detergents on experimental cholesterol atheromatosis of rabbits. Archives Pathol 38:381–391

Irvine EJ, Marshall JK 2000 Increased intestinal permeability precedes the onset of Crohn's disease in a subject with familial risk. Gastroenterology 119:1740–1744

Mercurius-Taylor IA, Jayaraj AP, Clark CG 1984 Is chronic detergent ingestion harmful to the gut? Br J Industrial Med 40:279–281

Animal models of intestinal inflammation: clues to the pathogenesis of inflammatory bowel disease

Fiona Powrie and Holm Uhlig[1]

Sir William Dunn School of Pathology, University of Oxford, South Parks Road, Oxford OX1 3RE, UK

Abstract. In the last decade a number of models of chronic intestinal inflammation have been described that resemble aspects of the pathology found in patients with inflammatory bowel disease. Several themes have emerged from these studies that are of relevance to the pathogenesis of inflammatory bowel disease. Firstly, intestinal inflammation is a consequence of an aberrant chronic immune response triggered by enteric bacteria. Both innate and adaptive immune mechanisms can cause colitis and in many models there is evidence of differential activation of T helper 1 (Th1)-type cells. Targeting the Th1 pathway prevents experimental colitis and there is also evidence that this may be useful in Crohn's disease. Secondly, specialized populations of regulatory T cells have been shown to prevent colitis and in some systems cure it, suggesting immune responses in the intestine are subject to dominant T cell-mediated control. Here we focus on new insights into the pathogenesis and regulation of intestinal inflammation as revealed by model systems and how these may be harnessed for the treatment of IBD.

2004 Inflammatory bowel disease — crossroads of microbes, epithelium and immune systems. Wiley, Chichester (Novartis Foundation Symposium 263) p 164–178

A large number and vast array of commensal bacteria colonize the gastrointestinal tract in humans. In general this is a symbiotic relationship with benefits for both host and bacteria. However evidence from both clinical and experimental studies suggests that aberrant inflammatory responses to resident bacteria may be involved in the pathogenesis of inflammatory bowel disease (IBD) (Bouma & Strober 2003).

Susceptibility to IBD is determined by interactions between an individual's genetic make-up and environmental factors (Podolsky 2002). The complexity of these interactions make it difficult to dissect the role of individual factors in disease

[1]Present address: Division of Gastroenterology, Children's Hospital, University of Leipzig, Germany.

pathogenesis. In the last decade a number of models of chronic intestinal inflammation in rodents have been described. Many of these share features with the human disease and have provided excellent tools to study how genetic factors influence the host immune response and intestinal epithelial barrier function.

Models of intestinal inflammation can be divided into four groups: spontaneously arising, genetically induced as a result of deletion or over-expression of genes, chemically induced and T cell transfer into immuno-deficient mice (for a detailed review of individual models see Bouma & Strober 2003, Sartor 2004). Results from these studies have shown that IBD can develop as a consequence of altered intestinal barrier function, excessive effector cell responses or a deficiency in regulatory pathways. Many different genetic lesions lead to diseases with similar histopathological features suggesting that common pathways mediate the intestinal inflammatory response.

Development of intestinal inflammation requires enteric commensal bacteria as genetically susceptible mice raised on a germ-free background do not develop IBD (Sartor 2004). Intestinal bacteria are capable of providing antigenic and adjuvant stimuli and may drive both the initiation and perpetuation of the inflammatory response. No specific bacterial pathogen or common antigens have been identified. Instead it appears that several different bacterial species can induce intestinal inflammation. This is influenced by the host as particular bacteria can induce disease in one model but not another. The finding that host genetic factors influence which bacteria drive the inflammatory response has important implications for IBD in humans and may contribute to the phenotypic heterogeneity of the disease.

In the majority of models immune pathology is mediated by differential activation of T helper 1 (Th1)-type $CD4^+$ T cells (Strober et al 2002). These cells secrete high levels of tumour necrosis factor (TNF)α and interferon (IFN)γ and neutralization of these cytokines prevents disease. This is particularly relevant to Crohn's disease which also involves a Th1 cell response (Pallone & Monteleone 1998) and has been shown to be responsive to anti-TNFα therapies (Targan 2000). Polarized Th2 responses can also lead to intestinal inflammation in some models and it has been suggested these may be involved in the pathogenesis of ulcerative colitis (Bouma & Strober 2003), although further work is required to substantiate these findings.

A theme that has emerged from a number of the animal models is that development of intestinal inflammation is a balance between pathological effector responses and regulatory mechanisms that can inhibit inflammation (Powrie 1995). Here, we focus on studies from this laboratory using the T cell transfer model of colitis that have revealed the role of distinct $CD4^+$ T cells in the development and regulation of intestinal inflammation.

Development of IBD by depletion of regulatory T cells

The T cell transfer model of colitis was first described over a decade ago and has since been used extensively to probe the role of different immune cells in the induction and regulation of intestinal inflammation (Singh et al 2001). Early studies showed that transfer of primarily naïve ($CD45RB^{high}$) $CD4^+$ T cells to immunodeficient recipients led to wasting disease and colitis (Morrissey et al 1993, Powrie et al 1993). Intestinal inflammation resembles that seen in IBD with transmural leukocytic infiltrates, epithelial cell hyperplasia and goblet cell depletion. Th1 cells accumulate in the colon and drive the pathogenesis of disease as treatments which inhibit the differentiation or effector function of Th1 cells such as administration of anti-IFNγ, anti-TNF or anti-IL12p40 monoclonal antibodies (mAbs) inhibit disease (Singh et al 2001). As in most models, the pathogenic Th1 response is driven by resident bacteria as colitis fails to develop in T cell restored immunodeficient recipients raised under germ free conditions (Singh et al 2001, Aranda et al 1997).

One of the key findings from the T cell transfer model is that functionally specialized regulatory T (T_{reg}) cells prevent the immune activation caused by transfer of $CD4^+CD45RB^{high}$ T cells. Initially identified as being contained within the antigen-experienced $CD4^+CD45RB^{low}$ population (Powrie et al 1993), T_{reg} cells have more recently been found to enrich within the naturally activated subset that express CD25 (Read et al 2000). $CD4^+CD25^+$ cells represent 5–10% of peripheral and thymic $CD4^+$ T cells and are present in mice, rats and humans (Maloy & Powrie 2001). First identified for their ability to suppress autoimmune disease (Sakaguchi et al 1995), $CD4^+CD25^+$ T_{reg} are now known to mediate a more general suppressive role inhibiting responses to both self and environmental antigens (Gavin & Rudensky 2003).

A feature that distinguishes $CD4^+CD25^+T_{reg}$ cells from other $CD4^+$ T cells is that they adopt their effector function in the thymus (Sakaguchi et al 2003). Studies using TCR transgenic mice suggest that their thymic selection requires high affinity interaction with self-antigen (Jordan et al 2001). Exciting recent studies showed that the forkhead winged-helix transcription factor, FoxP3 is differentially expressed by $CD4^+CD25^+T_{reg}$ cells and is required for their differentiation (Khattri et al 2003, Fontenot et al 2003). Retroviral transduction with FoxP3 led to the development of regulatory T cells from naïve precursors indicating FoxP3 may be a master control switch for T_{reg} cell development (Hori et al 2003). Although the relevance of these immunosuppressive T cells in human IBD has yet to be established, loss of function mutations in FoxP3 have been shown to be responsible for immune dysregulation, polyendocrinopathy, enteropathy and X-linked (IPEX) syndrome (Bennett et al 2001). Strikingly patients with IPEX develop a similar spectrum of diseases to those which develop in animal models with deficiencies in regulatory T cells including Type 1 diabetes, allergy and IBD-like enteropathy.

In addition to naturally occurring $CD4^+CD25^+T_{reg}$ cells there are also induced regulatory T cell populations that develop from naïve T cells following chronic antigen stimulation. Amongst these, Tr1 cells secrete high levels of interleukin (IL)10 (Groux et al 1997) and Th3 cells secrete TGFβ (Chen et al 1994, Strober et al 1997). Both populations have been shown to inhibit colitis via mechanisms involving immunosuppressive cytokines and may represent another layer of immune regulation which together with naturally occurring T_{reg} cells prevents immune pathology during chronic immune responses.

Role of immunosuppressive cytokines in the control of intestinal inflammation

Mice with genetically induced deletions of IL2 (Sadlack et al 1993), transforming growth factor (TGF)β (Kulkarni et al 1993) and IL10 (Kuhn et al 1993) all develop colitis indicating important roles for these cytokines in intestinal homeostasis. Inflammatory disease in *IL2*$^{-/-}$ mice is not restricted to the colon and these mice develop multi-organ inflammatory disease (Sadlack et al 1993). It is thought that systemic immune cell hyperactivity in *IL2*$^{-/-}$ mice is a consequence of a lack of $CD4^+CD25^+T_{reg}$ cells which require IL2 for their development in the thymus (Malek 2003).

Inflammation in *IL10*$^{-/-}$ mice is restricted to the colon suggesting that IL10 is not required for control of self-reactive T cells but plays a more important role in regulating responses to exogenous stimuli. Indeed *IL10*$^{-/-}$ mice mount hyperactive immune responses to a variety of infectious agents and inflammatory stimuli (Moore et al 2001). Direct evidence that IL10 produced in response to intestinal bacteria plays a key regulatory role comes from studies of *Helicobacter hepaticus* infection in mice. Despite its persistence in the intestine, *H. hepaticus* does not induce intestinal inflammation in wild-type mice but triggers a chronic Th1-mediated colitis in *IL10*$^{-/-}$ mice (Kullberg et al 2001). Protection in wild-type mice is associated with bacteria-induced IL10-secreting T_{reg} cells that are present in both $CD25^+CD45RB^{low}$ and $CD25^-CD45RB^{low}$ populations and prevent *H. hepaticus*-induced colitis in T cell-restored immunodeficient mice (Kullberg et al 2002).

In the T-cell transfer model of colitis, administration of anti-TGFβ (Powrie et al 1996) or anti-IL10R mAb (Asseman et al 1999) abrogated the ability of $CD4^+CD45RB^{low}$ cells to inhibit colitis indicating non-redundant roles for both cytokines in the control of intestinal inflammation. IL10 (Moore et al 2001) and TGFβ (Letterio & Roberts 1998) are pleiotropic cytokines capable of mediating inhibitory effects on a number of different cell types. In recent studies it was found that abrogation of TGFβ signalling in T cells was sufficient to induce colitis indicating that TGFβ acts directly on T cells to control intestinal homeostasis (Gorelik & Flavell 2000). In contrast, disruption of IL10 signalling in myeloid cells, and neutrophils but not T cells, triggered colitis suggesting that IL10 controls intestinal inflammation via inhibition of innate immune cell activation (Takeda et al 1999). The requirement for TGFβ and IL10 for T_{reg} activity may therefore reflect the action of these cytokines on different cell populations in the inflammatory cascade. It may also indicate that these two cytokines are linked and TGFβ has recently been shown to induce IL10 production (Kitani et al 2003).

Activated dendritic cells drive intestinal inflammation

Central to understanding the pathogenesis of chronic intestinal inflammation is the identification of antigen-presenting cells (APCs) in intestinal mucosa and secondary lymphoid organs that are involved in the initiation and perpetuation of the inflammatory response. Dendritic cells (DCs) have a unique capacity to stimulate naïve T cells and are therefore good candidates to drive intestinal inflammation. While there have been many studies on small intestinal DCs, much less is known about DCs in the colon. In the normal colon, DCs are scattered throughout the mucosa and are present in small clusters of leukocytic cells (Malmstrom et al 2001, Mottet et al 2003, Krajina et al 2003). DCs express on their surface a number of receptors, termed pattern recognition receptors (PRRs), capable of recognizing molecular structures on microbes. Signalling via PRRs activates DCs, which produce inflammatory cytokines and costimulatory molecules that are required for sustained T cell activation (Reis e Sousa 2004). Little is known about how DCs sense commensal bacteria and how this governs development of local T cell responses in the healthy and inflamed intestine. Recent studies have shown that intestinal DCs are able to interact directly with enteric bacteria via transepithelial dendrites (Rescigno et al 2001). Bacterial-induced activation of DCs is thought to enable them to migrate to the secondary lymphoid organs and activate naïve T cells with specificity for intestinal antigens (Neutra et al 1996).

In mice with colitis there is an accumulation of DCs in the mesenteric lymph nodes (MLNs) and colon. Colitis is also associated with a higher proportion of DCs in the MLNs expressing an activated phenotype with elevated expression of CD40 and CD134L (Malmstrom et al 2001, Singh et al 2001). The latter is a TNF family molecule induced on DCs following activation via CD40 (Brocker et al 1999). Its receptor, CD134 is expressed on activated T cells and delivers costimulatory signals involved in the maintenance of the T cell response (Lane 2000). $CD134^+$ DCs are involved in the development of colitis as administration of anti-CD134L mAbs prevents disease (Malmstrom et al 2001). Anti-CD154 mAbs also inhibit colitis indicating that interactions between $CD154^+$-activated T cells and $CD40^+$ DCs play a similarly important role in disease pathogenesis (Liu et al 2000, De Jong et al 2000). Altogether these data suggest that interactions between activated DCs and activated T cells drive intestinal inflammation and that disruption of this positive-feedback loop may be beneficial in the treatment of colitis.

Analysis of T cell proliferation in mice with colitis showed that up to 20–30% of T cells were proliferating in the MLNs which fits with the high frequency of activated DCs at this site (Mottet et al 2003). Somewhat surprisingly, a similar frequency of T cells also proliferated locally in the colon. Proliferating T cells were typically located adjacent to DCs in organized leukocyte clusters in the lamina propria, suggesting that these local structures may contribute to a sustained inflammatory response. T cells homed to DC clusters in the colon very early after transfer into immunodeficient recipients, prior to the onset of colitis, making it tempting to speculate that colonic leukocyte clusters may also play a role in the initiation of the immune response (Leithauser et al 2001).

Transfer of $CD4^+CD25^+$ T_{reg} cells prevented the accumulation of activated DC in immunodeficient mice transfused with $CD4^+CD45RB^{high}$ T cells suggesting that T_{reg} cells act to inhibit the ability of DCs to induce a sustained T cell response (Malmstrom et al 2001). Whether they do this via effects on DC migration or activation or both remains to be established.

Suppression of the innate immune response by $CD4^+CD25^+T_{reg}$ cells

In addition to inducing T cell-dependent colitis *H. hepaticus* infection induces T cell independent colitis in 129SvEv $RAG^{-/-}$ mice (Maloy et al 2003). Intestinal inflammation in the colon and the caecum is accompanied by sustained activation of the innate immune system both locally in the intestine and in the spleen. The proinflammatory cytokines IFNγ, TNFα and IL12p40 are involved in the inflammatory cascade as administration of neutralizing monoclonal antibodies reactive with any one of these cytokines was sufficient to inhibit disease (Maloy et al 2003).

Transfer of $CD4^+CD25^+T_{reg}$ cells to *H. hepaticus*-infected $RAG^{-/-}$ mice prevented intestinal inflammation and innate immune cell activation (Maloy et al 2003). T_{reg} cells were found to accumulate in the MLN and colon and suppressed innate immune pathology via IL10- and TGFβ-dependent mechanisms. $CD4^+CD25^+T_{reg}$ cells from *IL10*$^{-/-}$ mice failed to inhibit colitis suggesting that IL10 production is crucial for their function. Kinetic studies showed that the suppressive effects of $CD4^+CD25^+T_{reg}$ cells were most evident later in infection suggesting T_{reg} cells do not inhibit the initial activation of the innate immune response but act to prevent chronic activation (K. Maloy, unpublished). This makes teleological sense as it would be deleterious to host-protective responses if T_{reg} cells prevented initial activation of the innate immune response. Indeed, recent studies have shown that innate stimulation of DCs abrogates T_{reg} cell activity via a mechanism involving IL6 (Pasare & Medzhitov 2003).

To date little is known about the specificity of T_{reg} cells that control intestinal inflammation. $CD4^+CD25^+T_{reg}$ cells isolated from mice that had not been infected with *H. hepaticus* were capable of preventing T cell independent colitis indicating that T_{reg} cells can suppress immune pathology triggered by organisms to which they have not been exposed (Maloy et al 2003). However the ability to control T cell dependent colitis was enhanced in T_{reg} populations from *H. hepaticus*-infected donors suggesting that bacterial infection can drive antigen-specific T_{reg} cells (Kullberg et al 2002).

The ability of $CD4^+CD25^+T_{reg}$ cells to control the innate immune response may underlie the ability of these cells to inhibit chronic immune responses to a number of infectious agents and represent an important host mechanism to prevent immune pathology in the face of chronic immune stimulation (Hori et al 2002, Maloy et al 2003, Belkaid et al 2002).

Cure of colitis by $CD4^+CD25^+T_{reg}$ cells

The ability of various populations of T_{reg} cells to prevent development of experimental colitis suggests that stimulation of these cells may be beneficial in patients with IBD. However, to be of use in the clinical setting T_{reg} cells must be able to inhibit ongoing T cell responses and reverse established pathology. In recent studies we found that a single transfusion of $CD4^+CD25^+T_{reg}$ cells to mice with established colitis led to resolution of the inflammatory response and restoration of normal intestinal architecture (Mottet et al 2003). Others found similar results and showed furthermore that the therapeutic effect of $CD4^+CD25^+T_{reg}$ cells is mediated via IL10 and TGFβ (Liu et al 2003) .

A characteristic feature of $CD4^+CD25^+$ cells is their inability to undergo mitogen or antigen-induced proliferation *in vitro* (Shevach 2002). By contrast, after transfer into colitic mice, T_{reg} cells were found to proliferate in MLN and colon where they inhibited the proliferation of pathogenic $CD4^+$ T cells (Mottet et al 2003). The presence of regulatory T cells in the colon was also associated with cure of colitis by Tr1 cells (Foussat et al 2003). Both studies would suggest that suppression of established inflammation requires T_{reg} activity in the inflamed effector sites. Proliferation of $CD4^+CD25^+T_{reg}$ cells was significantly reduced following resolution of the inflammatory response (Mottet et al 2003) suggesting that T_{reg} cells may respond to the inflammation that they regulate.

In the MLN and colon $CD4^+CD25^+T_{reg}$ localized at the interface between pathogenic effector T cells and DCs, and were in direct contact with both populations of cells (Mottet et al 2003). This specific distribution *in vivo* may reflect the ability of activated DCs to induce the activation and migration of $CD4^+CD25^+T_{reg}$ cells (Cederbom et al 2000, Oldenhove et al 2003, Yamazaki et al 2003, Bystry et al 2001). T_{reg} cells in turn may suppress the immune response via effects on activated DCs (Oldenhove et al 2003) or via direct effects on effector T cells (Piccirillo & Shevach 2001).

Concluding remarks

Animal models of intestinal inflammation have provided excellent tools to dissect the molecular and cellular pathways involved in the development and control of intestinal inflammation. Key cytokines involved in the effector cell response have been identified providing a number of new therapeutic targets. Some of these, such as inhibition of TNFα, are already showing promising results. Development of intestinal inflammation was found to be a balance between effector and regulatory pathways. Transfer of T_{reg} cells was sufficient to cure experimental colitis suggesting that enhancement of T_{reg} cell activity may be of benefit for the treatment of IBD.

Acknowledgements

F. Powrie is funded by the Wellcome Trust and Holm Uhlig by the EU.

References

Aranda R, Sydora BC, McAllister PL et al 1997 Analysis of intestinal lymphocytes in mouse colitis mediated by transfer of CD4+, CD45RBhigh T cells to SCID recipients. J Immunol 158:3464–3473

Asseman C, Mauze S, Leach MW, Coffman RL, Powrie F 1999 An essential role for interleukin 10 in the function of regulatory T cells that inhibit intestinal inflammation. J Exp Med 190:995–1004

Belkaid Y, Piccirillo CA, Mendez S, Shevach EM, Sacks DL 2002 CD4+CD25+ regulatory T cells control Leishmania major persistence and immunity. Nature 420:502–507

Bennett CL, Christie J, Ramsdell F et al 2001 The immune dysregulation, polyendocrinopathy, enteropathy, X-linked syndrome (IPEX) is caused by mutations of FOXP3. Nat Genet 27: 20–21

Bouma G, Strober W 2003 The immunological and genetic basis of inflammatory bowel disease. Nat Rev Immunol 3:521–533

Brocker T, Gulbranson-Judge A, Flynn S et al 1999 CD4 T cell traffic control: in vivo evidence that ligation of OX40 on CD4 T cells by OX40-ligand expressed on dendritic cells leads to the accumulation of CD4 T cells in B follicles. Eur J Immunol 29:1610–1616

Bystry RS, Aluvihare V, Welch KA, Kallikourdis M, Betz AG 2001 B cells and professional APCs recruit regulatory T cells via CCL4. Nat Immunol 2:1126–1132

Cederbom L, Hall H, Ivars F 2000 CD4+CD25+ regulatory T cells down-regulate co-stimulatory molecules on antigen-presenting cells. Eur J Immunol 30:1538–1543

Chen Y, Kuchroo VK, Inobe J, Hafler DA, Weiner HL 1994 Regulatory T cell clones induced by oral tolerance: suppression of autoimmune encephalomyelitis. Science 265:1237–1240

De Jong YP, Comiskey M, Kalled SL et al 2000 Chronic murine colitis is dependent on the CD154/CD40 pathway and can be attenuated by anti-CD154 administration. Gastroenterology 119:715–723
Fontenot JD, Gavin MA, Rudensky AY 2003 Foxp3 programs the development and function of CD4+CD25+ regulatory T cells. Nat Immunol 4:330–336
Foussat A, Cottrez F, Brun V et al 2003 A comparative study between T regulatory type 1 and CD4+CD25+ T cells in the control of inflammation. J Immunol 171:5018–5026
Gavin M, Rudensky A 2003 Control of immune homeostasis by naturally arising regulatory CD4+ T cells. Curr Opin Immunol 15:690–696
Gorelik L, Flavell RA 2000 Abrogation of TGFbeta signaling in T cells leads to spontaneous T cell differentiation and autoimmune disease. Immunity 12:171–181
Groux H, O'Garra A, Bigler M et al 1997 A CD4+ T-cell subset inhibits antigen-specific T-cell responses and prevents colitis. Nature 389:737–742
Hori S, Carvalho TL, Demengeot J 2002 CD25+CD4+ regulatory T cells suppress CD4+ T cell-mediated pulmonary hyperinflammation driven by Pneumocystis carinii in immunodeficient mice. Eur J Immunol 32:1282–1291
Hori S, Nomura T, Sakaguchi S 2003 Control of regulatory T cell development by the transcription factor Foxp3. Science 299:1057–1061
Jordan MS, Boesteanu A, Reed AJ et al 2001 Thymic selection of CD4+CD25+ regulatory T cells induced by an agonist self-peptide. Nat Immunol 2:301–306
Khattri R, Cox T, Yasayko SA, Ramsdell F 2003 An essential role for Scurfin in CD4+CD25+ T regulatory cells. Nat Immunol 4:337–342
Kitani A, Fuss I, Nakamura K et al 2003 Transforming growth factor (TGF)-beta1-producing regulatory T cells induce Smad-mediated interleukin 10 secretion that facilitates coordinated immunoregulatory activity and amelioration of TGF-beta1-mediated fibrosis. J Exp Med 198:1179–1188
Krajina T, Leithauser F, Moller P, Trobonjaca Z, Reimann J 2003 Colonic lamina propria dendritic cells in mice with CD4+ T cell-induced colitis. Eur J Immunol 33: 1073–1083
Kuhn R, Lohler J, Rennick D, Rajewsky K, Muller W 1993 Interleukin-10-deficient mice develop chronic enterocolitis. Cell 75:263–274
Kulkarni AB, Huh CG, Becker D et al 1993 Transforming growth factor beta 1 null mutation in mice causes excessive inflammatory response and early death. Proc Natl Acad Sci USA 90: 770–774
Kullberg MC, Rothfuchs AG, Jankovic D et al 2001 Helicobacter hepaticus-induced colitis in interleukin-10-deficient mice: cytokine requirements for the induction and maintenance of intestinal inflammation. Infect Immun 69:4232–4241
Kullberg MC, Jankovic D, Gorelick PL et al 2002 Bacteria-triggered CD4(+) T regulatory cells suppress Helicobacter hepaticus-induced colitis. J Exp Med 196:505–515
Lane P 2000 Role of OX40 signals in coordinating CD4 T cell selection, migration, and cytokine differentiation in T helper (Th)1 and Th2 cells. J Exp Med 191:201–206
Leithauser F, Trobonjaca Z, Moller P, Reimann J 2001 Clustering of colonic lamina propria CD4(+) T cells to subepithelial dendritic cell aggregates precedes the development of colitis in a murine adoptive transfer model. Lab Invest 81:1339–1349
Letterio JJ, Roberts AB 1998 Regulation of immune responses by TGF-beta. Annu Rev Immunol 16:137–161
Liu H, Hu B, Xu D, Liew FY 2003 CD4+CD25+ regulatory T cells cure murine colitis: the role of IL-10, TGF-beta, and CTLA4. J Immunol 171:5012–5017
Liu Z, Geboes K, Colpaert S et al 2000 Prevention of experimental colitis in SCID mice reconstituted with CD45RBhigh CD4+ T cells by blocking the CD40-CD154 interactions. J Immunol 164:6005–6014

Malek TR 2003 The main function of IL-2 is to promote the development of T regulatory cells. J Leukoc Biol 74:961–965
Malmstrom V, Shipton D, Singh B et al 2001 CD134L expression on dendritic cells in the mesenteric lymph nodes drives colitis in T cell-restored SCID mice. J Immunol 166: 6972–6981
Maloy KJ, Powrie F 2001 Regulatory T cells in the control of immune pathology. Nat Immunol 2:816–822
Maloy KJ, Salaun L, Cahill R et al 2003 CD4+CD25+ T(R) cells suppress innate immune pathology through cytokine-dependent mechanisms. J Exp Med 197:111–119
Moore KW, de Waal Malefyt R, Coffman RL, O'Garra A 2001 Interleukin-10 and the interleukin-10 receptor. Annu Rev Immunol 19:683–765
Morrissey PJ, Charrier K, Braddy S, Liggitt D, Watson JD 1993 CD4+ T cells that express high levels of CD45RB induce wasting disease when transferred into congenic severe combined immunodeficient mice. Disease development is prevented by cotransfer of purified CD4+ T cells. J Exp Med 178:237–244
Mottet C, Uhlig HH, Powrie F 2003 Cutting edge: cure of colitis by CD4+CD25+ regulatory T cells. J Immunol 170:3939–3943
Neutra MR, Pringault E, Kraehenbuhl JP 1996 Antigen sampling across epithelial barriers and induction of mucosal immune responses. Annu Rev Immunol 14:275–300
Oldenhove G, de Heusch M, Urbain-Vansanten G et al 2003 CD4+ CD25+ regulatory T cells control T helper cell type 1 responses to foreign antigens induced by mature dendritic cells in vivo. J Exp Med 198:259–266
Pallone F, Menteleone G 1998 Interleukin 12 and Th1 responses in inflammatory bowel disease. Gut 43:735–736
Pasare C, Medzhitov R 2003 Toll pathway-dependent blockade of CD4+CD25+ T cell-mediated suppression by dendritic cells. Science 299:1033–1036
Piccirillo CA, Shevach EM 2001 Cutting edge: control of CD8+ T cell activation by CD4+ CD25+ immunoregulatory cells. J Immunol 167:1137–1140
Podolsky DK 2002 Inflammatory bowel disease. N Engl J Med 347:417–429
Powrie F 1995 T cells in inflammatory bowel disease: protective and pathogenic roles. Immunity 3:171–174
Powrie F, Leach MW, Mauze S, Caddle LB, Coffman RL 1993 Phenotypically distinct subsets of CD4+ T cells induce or protect from chronic intestinal inflammation in C. B-17 scid mice. Int Immunol 5:1461–1471
Powrie F, Carlino J, Leach MW, Mauze S, Coffman RL 1996 A critical role for transforming growth factor-beta but not interleukin 4 in the suppression of T helper type 1-mediated colitis by CD45RB(low) CD4+ T cells. J Exp Med 183:2669–2674
Read S, Malmstrom V, Powrie F 2000 Cytotoxic T lymphocyte-associated antigen 4 plays an essential role in the function of CD25(+)CD4(+) regulatory cells that control intestinal inflammation. J Exp Med 192:295–302
Reis e Sousa C 2004 Activation of dendritic cells: translating innate into adaptive immunity. Curr Opin Immunol 16:21–25
Rescigno M, Urbano M, Valzasina B et al 2001 Dendritic cells express tight junction proteins and penetrate gut epithelial monolayers to sample bacteria. Nat Immunol 2:361–367
Sadlack B, Merz H, Schorle H et al 1993 Ulcerative colitis-like disease in mice with a disrupted interleukin-2 gene. Cell 75:253–261
Sakaguchi S, Sakaguchi N, Asano M, Itoh M, Toda M 1995 Immunologic self-tolerance maintained by activated T cells expressing IL-2 receptor alpha-chains (CD25). Breakdown of a single mechanism of self-tolerance causes various autoimmune diseases. J Immunol 155:1151–1164

Sakaguchi S, Hori S, Fukui Y et al 2003 Thymic generation and selection of CD25+CD4+ regulatory T cells: implications of their broad repertoire and high self-reactivity for the maintenance of immunological self-tolerance. In: Generation and regulatory functions of lymphocytes (Novartis Found Symp 252) p 6–23

Sartor RB 2004 Therapeutic manipulation of the enteric microflora in inflammatory bowel diseases: antibiotics, probiotics, and prebiotics. Gastroenterology 126:1620–1633

Shevach EM 2002 CD4+ CD25+ suppressor T cells: more questions than answers. Nat Rev Immunol 2:389–400

Singh B, Read S, Asseman C et al 2001 Control of intestinal inflammation by regulatory T cells. Immunol Rev 182:190–200

Strober W, Kelsall B, Fuss I et al 1997 Reciprocal IFN-gamma and TGF-beta responses regulate the occurrence of mucosal inflammation. Immunol Today 18:61–64

Strober W, Fuss IJ, Blumberg RS 2002 The immunology of mucosal models of inflammation. Annu Rev Immunol 20:495–549

Targan SR 2000 Biology of inflammation in Crohn's disease: mechanisms of action of anti-TNFα therapy. Can J Gastroenterol 14(suppl C): 13C–16C

Takeda K, Clausen BE, Kaisho T et al 1999 Enhanced Th1 activity and development of chronic enterocolitis in mice devoid of Stat3 in macrophages and neutrophils. Immunity 10:39–49

Yamazaki S, Iyoda T, Tarbell K et al 2003 Direct expansion of functional CD25+ CD4+ regulatory T cells by antigen-processing dendritic cells. J Exp Med 198:235–247

DISCUSSION

Ghosh: T lymphocytes don't normally express much in the way of Toll receptors. Do regulatory T cells express more Toll receptors, and what is the controversy here?

Powrie: Apparently there are now convincing reports that some T cells do express Toll receptors. There was a report that $CD25^+$ cells express TLR4 and 7 with some proliferative response to LPS (Caramalho et al 2003). I haven't seen this confirmed, but if they are able to respond to inflammation in this way this would be important.

Sartor: I have a question concerning the antigen specificity of the $CD25^+$ cells. It has been suggested that unless you obtain the $CD4^+$ regulatory cell population from an animal infected with *Helicobacter* you don't get protection. This suggests indirectly that they had antigen specificity for that organism. Yet these are thymic self-antigen selected. What is your opinion on the antigen specificity and did you find it was a *Helicobacter*-specific donor?

Powrie: I didn't have time to describe all the complexity. Different subsets of $CD4^+$ cells can regulate in this model. Cells within the $CD45RB^{low}CD25^-$ pool appear to be sensitive to exposure to *H. hepaticus* in the donor. Over the doses of cells that we have tested we have not been able to reveal an effect of priming for the $CD25^+$ compartment in that we take these cells from mice that are not infected with *Helicobacter*. This would suggest it is not a classic memory cell response. However, we have to do further dose response experiments to be sure of this. Of course, we know that TCR receptor recognition is degenerate and one TCR can recognize up to 10^6 different peptides. It is quite possible that $CD4^+CD25^+$ T_{reg} cells that are selected on the basis of self-reactivity will cross react with microbial components. Obviously the specificity of the $CD4^+CD25^+$ T_{reg} cells that control *H. hepaticus*-driven inflammation remains a key question.

Sartor: I have a slightly heretical thought. You have clearly shown that IL10 is important in the regulatory cell activity. What is your opinion about the source of the IL10: are you convinced that it is necessary from the T cell rather than the APC? And could one interpret the failure of an IL10-positive cell to induce disease to be a result of the lack of donor activation due to a lack of appropriate stimulation of a regulatory pathway?

Powrie: These are important issues. As far as I am aware there are no mice with cell-type specific deletions of IL10 available that will allow us to address some of these things. I am not saying T_{reg} cell-produced IL10 is the sole source of IL10, only that it is an important one. Host-derived IL10 might play an equally important role.

Sartor: In our mix and match situation if we take APCs from an IL10 knockout they are normal and T cells from an IL10 knockout are normal. *In vitro* we can get activation of IL10 knockout T cells to produce IFNγ in response to bacteria when they are co-cultured with IL10 knockout APCs but not vice versa.

Powrie: Clearly myeloid cell-derived IL10 plays an important role in that type of situation. This hasn't been addressed in the model systems. We have to go to the next stage of looking at the role of other cell types.

Sartor: DC transfers could address that as well.

Schölmerich: Florian Obermeier from our group did an unusual experiment. He treated the donor mice in this model with bacterial DNA fragments (CpG motifs). The T cells from the treated donors were not able to induce colitis in the SCID mice in the same way that T cells from untreated donors could. Is there an explanation for this? The donor is given a non-specific immune stimulation that prevents induction of colitis in the recipient.

Powrie: I don't know, unless you had some deletion of pathogenic cells. There is a balance of effector cells versus regulation. One would have to look at what happens to the T cell subsets.

Schölmerich: Normally these DNA fragments would cause induction because they are stimulating the immune system, and not deletion.

Powrie: You have to look at the phenotypes of the cells and what happens in different populations.

Moore: Have you tried the anti-CD134 ligand antibody therapeutically?

Powrie: Yes, and it doesn't work. The only thing we have found to be convincing with regards to therapy is the transfer of regulatory T cells. This is a very aggressive model with severe intestinal inflammation. We have even combined anti-CD134L with other approaches. This is why I feel we must reset the immunological balance when we talk about manipulating the immune system in a chronic disease. It is not sufficient to simply block the effector cells.

Moore: Have you ever taken your cells out of your SCID mouse later on and had a look at the phenotype? At the start you are co-injecting with the high and low $CD25^+$: what have you got at the end?

Powrie: In the experiments I showed we used genetically marked cells, so we can pull them out and look at them. We have looked at IL10 secretion in the progeny of $CD25^+$ cells as well as the Rb^{high} progeny. We don't see a large induction of IL10 among the Rb^{high} progeny, but we do see IL10 secretion in certain sites by the progeny of the $CD25^+$ population.

Jewell: Do you want to say more about how the T_{reg} cells recognize the activated DCs? Is it a CD40/CD40L interaction? You said it was not antigen specific.

Powrie: I am not saying it isn't antigen specific, we just don't know the antigens. The concept here is that these cells can mediate antigen-driven bystander suppression, so when they see their antigen and they are activated, these cells are then able to turn off responses to other antigens. This is what you want in an inflammatory lesion where we don't know what antigens are driving the response. We need to get cells in, and they need to be activated by an antigen—they may be self-reactive or cross-reactive with microbial antigens. This then activates them to mediate their suppressor/effector functions. What is really important is how these cells get to these places and how they know to talk to activated DCs. The chemokine/chemokine receptor side is not very well characterized.

Mahida: From your paper and the subsequent discussion, am I right in thinking that you believe IL10 perhaps has a greater role than TGFβ in T_{reg}-mediated suppression?

Powrie: This is another issue that divides the T_{reg} community. In our hands IL10 and TGFβ both appear to be required. If you block either of these then you abrogate T_{reg} function. Both of these cytokines have multiple effects on different populations. There are some nice data from Warren Strober's group showing that TGFβ can up-regulate IL10, indicating that there are connections between these cytokines at the molecular level. To get on top of established intestinal inflammation you may need both.

Mahida: In your Rag knockout experiments how do you think the T_{reg} cells are mediating the so-called anti-inflammatory effect?

Powrie: Kevin Maloy's data provided a new idea on how $CD25^+$ cells modulate the inflammatory response. Previously, $CD4^+CD25^+$ T_{reg} were thought to work primarily on other T cells, and often involving an uncharacterized cell contact-dependent mechanism. He showed that they can also have activities on the innate immune system. This is completely compatible with the functions of IL10 and

TGFβ. Precisely which cells are responding is unclear. It is probably something fairly early in the inflammatory cascade that then stops the elaboration of the whole thing.

Sartor: To amplify the necessity of IL10 and TGFβ, with Warren Strober we are using his nasal DNA probes. We have shown that there is no protection with TGFβ nasal plasmids in the IL10 knockout mouse, despite the production of TGFβ. We need to confirm this because we have so far done it three times but once it didn't work technically.

Powrie: That is a nice result.

Sartor: Work with IL10-transfected T cells and local gut delivery would suggest that IL10 can work at the gut level. It remains to be determined whether it would work in the absence of TGFβ.

Gibson: As a simple clinician, it seems pretty straightforward: all you need to do is give people T_{reg} cells. How do you do this, what strategy do you use to produce the cells, and if you could do this what other effects would there be?

Powrie: We don't know the answers. As a proof of principle this would be very important. IBD is one of the diseases that might be a good one to try; you have the inflammation and the cells should in theory home into these sites. Naturally occurring T_{reg} cells are subject to very tight homeostatic control and this may limit their activities to the inflammatory site. Moving this from mouse models into humans is attractive but very difficult. Some studies have been performed with an anti-CD3 antibody in Type 1 diabetes which seems to lead to the emergence of these $CD25^+$ cells, and this is a TGFβ-dependent effect.

Gibson: If you could do this, do you think there would be effects other than on the inflammation? Are we going to get fibrotic lungs, for example?

Powrie: I think this is why IL10 and TGFβ are both needed; they seem to balance each other. This needs to be looked at. When we do these sorts of transfers we need to look at what happens to normal immune responses to other things in the mice we are working with.

Schölmerich: We have isolated human regulatory T cells but we don't know what to do with them. They work nicely *in vitro* but we don't dare use them in people.

Powrie: Perhaps we need to develop methods to induce T_{reg} cell function *in situ*. We need to know about whether any of the IBD mutations affect this pathway.

Wright: In the models you have described trying to produce a fibrogenic reaction is difficult. Why is that? Do you know any dodges to increase the fibrosis?

Powrie: No.

Sartor: If you do it chronically and it is T cell mediated you do get some fibrosis. It depends on the insult. With the peptidoglycan model in the Lewis rat you get a lot of fibrosis.

Wright: I understand there is going to be a pan-European trial of GM-CSF-mobilized stem cells in severe Crohn's disease. There is anecdotal evidence that if you do this in selected patients with high Crohn's disease indices, their indices can be reduced. What is happening there?

Powrie: That is interesting but I don't really know how to interpret it.

Wright: They are using GM-CSF mobilized stem cells, harvesting them and putting them back intravenously. There is a recent paper (Craig et al 2003) showing that after autologous bone marrow transplantation Crohn's disease indices come right down. I am trying to understand why this happens.

Van Deventer: There have only been three cases reported. It is not only GM-CSF but also the cyclophosphamide that is always included. It probably isn't necessary to give the stem cells back. The trial in Europe will be set up so that there is only a moblilization/induction regimen with cyclophosphamide and GM-CSF in one arm and stem cells not given back, and in the other arm they will be given back. More than 500 patients have undergone an autologous stem cell transplant for other indications in Europe, and nobody really knows why it works. However, one of the things that is seen is that there is a long-term loss of effector T cells. There are data in rheumatoid arthritis to indicate that there might be something happening on the T_{reg} level as well.

Wright: One of the reasons I ask is that Tom McDonald and Mairi Brittan have shown in TNBS colitis that if you do mixed-sex bone marrow transplantation, 80% of the myofibroblasts you get are bone-marrow derived. You also get vasculogenesis with muscle cells and Rouget cells of bone marrow derivation. I would be interested to see what you find if you get tissue from these patients.

References

Caramalho I, Lopes-Carvalho T, Ostler D, Zelenay S, Haury M, Demengeot J 2003 Regulatory T cells selectively express toll-like receptors and are activated by lipopolysaccharide. J Exp Med 197:403–411

Craig RM, Traynor A, Oyama Y, Burt RK 2003 Hematopoietic stem cell transplantation for severe Crohn's disease. Bone Marrow Transplant 32(suppl 1):S57–59

Homing of intestinal immune cells

Holm H. Uhlig[1], Christian Mottet[2] and Fiona Powrie[3]

Sir William Dunn School of Pathology, University of Oxford, South Parks Road, Oxford OX1 3RE, UK

Abstract. The homing of immune cells into the intestinal mucosa, the gut-associated lymphoid tissue or the mesenteric lymph nodes involves a complex process of molecular events that is dependent on cell type and cell maturation. Key factors that collectively determine the homing of leukocytes and their interaction with resident endothelial, epithelial, stromal and immune cells are interactions between integrins or selectins with their tissue adhesion molecules as well as chemokine receptors and their ligands. The organization of the small and large intestinal tissue and the mucosa associated lymphoid tissue as well as the presence or absence of inflammatory stimuli influence the homing of intestinal immune cells. The homing pattern of intestinal dendritic cells and CD4$^+$ T cells and its role for the pathogenesis and regulation of inflammatory bowel disease are discussed.

2004 Inflammatory bowel disease — crossroads of microbes, epithelium and immune systems. Wiley, Chichester (Novartis Foundation Symposium 263) p 179–193

Homing as a complex process depending on cell type, tissue organization and inflammatory stimuli

A characteristic feature of the intestinal mucosal immune system is its close association with the epithelium and the intestinal lumen that contains large amounts of dietary, bacterial and viral antigens. Due to the pathogenic potential of different components in the food and the intestinal bacterial flora the intestinal immune system needs to be tightly regulated. It must allow rapid and sufficient immune response against pathogens but avoid extensive immune response towards dietary antigens, the commensal flora or autoantigens. The development of intestinal immune responses is based on a constant traffic of leukocytes between bone marrow, gut and lymphoid organs via blood and lymph.

---

[1]Present address: Division of Gastroenterology, Children's Hospital, University of Leipzig, Germany.

[2]Present address: Division of Gastroenterology & Hepatology, Universital Hospital Lausanne, Switzerland.

[3]This paper was presented at the symposium by Holm H. Uhlig. Correspondence should be addressed to Fiona Powrie.

The homing of immune cells to the intestinal mucosa, the mucosa-associated lymphoid tissue (MALT) or the mesenteric lymph node involves a complex process of molecular events (Kunkel et al 2003). General factors that influence the homing of intestinal leukocytes at different stages of their life cycle include interactions with resident endothelial, epithelial, and stromal cells as well as with leukocytes (Campbell et al 2003, Kunkel et al 2003). The cellular organization of the small and large intestinal mucosa and the presence of the MALT determine the homing (Kunkel et al 2003, Mowat 2003). During tissue inflammation additional tissue factors are secreted that modify the homing pattern of leukocytes.

Common steps in the homing of bone-marrow derived immune cells from the blood into their target organ include attachment to and rolling on the endothelial cell surface, leukocyte activation, arrest and diapedesis (Campbell et al 2003). This process is mediated by initial low affinity interactions of integrins (such as the $\alpha 4\beta 7$ integrin) and/or selectins (such as L-selectin, CD62L) on the surface of leukocytes with intestinal tissue specific adhesion molecules (such as the mucosal adhesion molecule, MAdCAM) on endothelial cells (Butcher & Picker 1996, Campbell et al 2003). This binding results in slowing of leukocyte movement, and allows the interaction of tissue- and inflammation-specific chemokines with chemokine receptors on the leukocyte surface. Chemokine receptors are G protein coupled and mediate cell activation signals, which include conversion of the binding of leukocyte integrins with tissue adhesion molecules to high affinity interactions. This induces cell arrest and subsequent migration through the epithelium and into the tissue (Campbell et al 2003). The chemokine receptors have recently been classified into four groups as CCR, CXCR, CX3CR and XCR chemokine receptors depending on the arrangement of the N-terminal cysteines (Gao & Metz 2003). The respective ligands are accordingly classified to the receptor nomenclature as CCL, CXCL, CX3CL and XCL chemokine receptor ligands (Gao & Metz 2003). The expression of various chemokine receptors on different cells and the promiscuous binding of several chemokines to chemokine receptors allows the precise modulation of immune cell homing via combinatory and sequential usage of ligands and receptors (Kunkel et al 2003). In the following sections, mechanisms will be highlighted that contribute to the homing of intestinal dendritic and $CD4^+$ T cells. The *in vivo* relevance of this homing for the development of the MALT, immune homeostasis and chronic intestinal inflammation will be discussed.

Homing of immune cells during the development of Peyer's patches

Whereas a large part of the intestinal lamina propria (LP) can be regarded as an effector site the intestinal immune system also contains organized lymphoid tissue to allow initiation of adaptive immune responses in direct proximity to luminal antigens (Mowat 2003). The involvement of different intestinal immune cells in the induction and/or effector phase of immune responses is reflected by differential homing towards the mucosa or the lymphoid tissue during different stages of their life cycle (Kunkel et al 2003). The organized lymphoid tissue in the mucosa comprises large aggregated lymphoid follicles such as Peyer's, caecal or appendix vermiformis associated patches (Mowat 2003). Furthermore there are non-aggregated lymphoid clusters defined as isolated lymphoid follicles (ILF) and cryptopatches (Mowat 2003, Kanamori et al 1996).

Peyer's patches (PPs) are well defined representatives of the MALT that have an important function in antigen uptake, antigen presentation and priming of mucosal immune responses (Kraehenbuhl & Neutra 2000, Shreedhar et al 2003, Mowat 2003). These lymphoid structures have a highly organized composition containing a specific M cell rich and goblet cell depleted follicle-associated epithelium (FAE) (Kraehenbuhl & Neutra 2000), a subepithelial dome (SED) region enriched in dendritic cells (DCs) (Iwasaki & Kelsall 2000) as well as B and T cell areas (Mowat 2003). The PPs contains three types of DC (Iwasaki & Kelsall 2001). There are $CD11b^+$ in the subepithelial dome, $CD8\alpha^+$ DCs in follicular areas and $CD11b^-CD8\alpha^-$ DCs in both places.

The development of PPs depends on a sequence of homing signals to allow the attraction of precursor and mature immune cells into the PP anlagen (Fig. 1). Cell–cell signals delivered via integrins or chemokines and their receptors are crucial events in this process. The organization of PPs starts with lymphatic vasculogenesis indicated by the accumulation and clustering of VCAM-1^+ stromal cells at embryonic day 15 (Yoshida et al 2001). This is followed by the attraction of bone marrow derived $CD3^-CD4^+$ $\alpha4\beta7$ integrin$^+$ progenitor cells that seed preformed VCAM-1^+ and MAdCAM-1^+ cell clusters at embryonic day 17.5. $CD3^-CD4^+CD45^+$ progenitor cells express $LT\alpha1\beta2$, and activate LT-βR expressing stromal cells in the PP anlagen (Mebius 2003). The NF-κB-inducing kinase is induced by LT-βR signalling leading to transcriptional control of CCL19, CCL21 and CXCL13. The relevance of this pathway is illustrated by the finding that NF-κB-inducing kinase-deficient mice and alymphoplasia mice with a spontaneous point mutation within this enzyme lack developed lymph nodes as well as PPs (Mebius 2003). LT-βR signalling also activates an alternative NF-κB pathway which is shared with the TNFRp55 and controls the expression of chemokines such as VCAM-1 (Mebius 2003). As a consequence of lymphotoxin (LT) signalling, further $\alpha4\beta1^+$ and $CXCR5^+$ cells are attracted into the PP anlagen. At this stage, $CXCR5^+$ $B220^+$ B lymphocytes and $CCR7^+$ T cells home to the PPs initiating the cellular organization into functionally distinct B and T cell areas. This final phase of B and T cell seeding starts at birth. The intestinal flora that colonize the intestine shortly after birth play an important role in the development of the PP structures acting as an environmental factor.

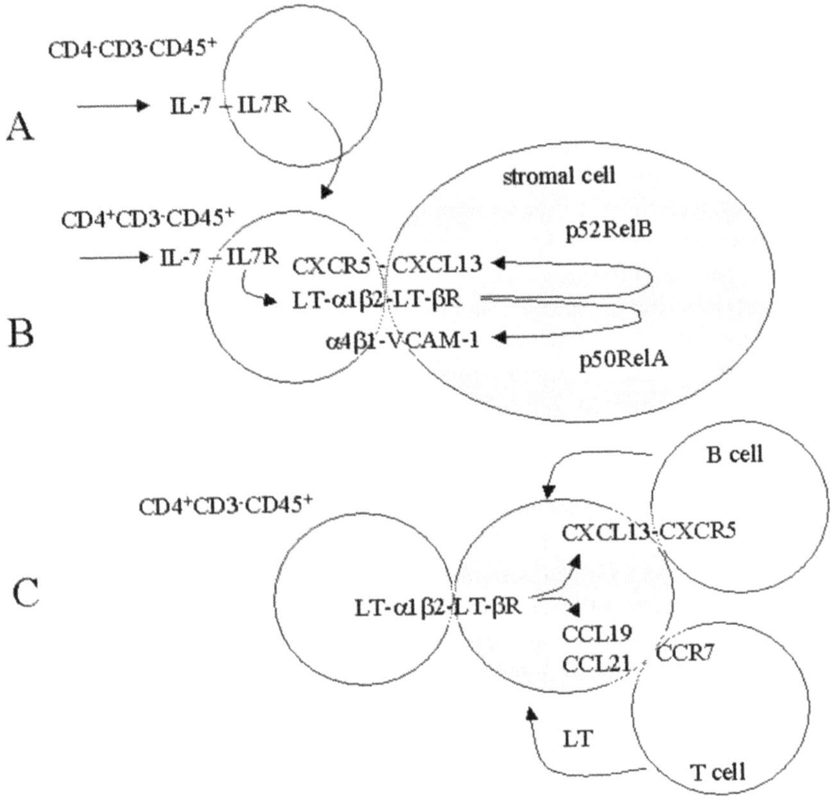

FIG. 1. Cellular and molecular interactions involved in the development of PPs. $IL7R^+$ $CD4^-CD3^-CD45^+$ precursor cells differentiate under the influence of IL7 into $CD4^+CD3^-$ cells (A) $CD4^+CD3^-$ $CXCR13^+\alpha 4\beta 7integrin^+$ cells home towards $MAdCAM^+$ endothelial and $CXCR5^+$ stromal cells in the PP anlagen (B). The interaction with stromal cells and continuous exposure to IL7 leads to LT-$\alpha 1\beta 2$ expression by $CD4^+CD3^-$ and allows the activation of LT-βR^+ stromal cells. LT-βR signalling induces further VCAM-1 and CXCR5 gene expression via p50RelA- and p52RelB-dependent signal cascades. This autocrine stimulation of $CD4^+CD3^-$ and stromal cells allows the expansion of $CD4^+CD3^-$/stromal cell clusters and the subsequent attraction of B and T cells (C). Under the influence of CXCL13 and CCL19 and CCL21, PP anlagen become organized into T and B cell areas.

Thus, germ free mice have a reduced number and size of PPs (Neutra et al 1996, Cebra 1999). Furthermore, the structure of the PP is disturbed resulting in a reduced number of T and B cells in both the follicles and in the FAE (Yamanaka et al 2003). Germ-free mice also fail to develop a germinal centre in the MALT (Cebra 1999).

Homing of intestinal DCs and CD4$^+$ T cells under non inflammatory conditions

The priming of intestinal CD4$^+$ T cells requires the interaction of naïve T cells with dendritic cells that present antigen via MHC-II and provide relevant costimulation within lymphoid tissues. Chemokine receptors, pathogen encounter, and the presence or absence of intestinal inflammation determine the maturation and activation state of DCs and their homing from the intestine into the lymphoid tissue where they interact with naïve CD4$^+$ T cells.

Bone marrow-derived DC precursors and monocytes home via the blood into the LP as well as to the subepithelial dome region of PPs. Under physiological steady-state conditions DCs are present in moderate density in the small intestine and scattered in low density within the colonic mucosa (Becker et al 2003, Malmstrom et al 2001, Krajina et al 2003). By contrast, DCs are found in high density within lymphoid follicles such as the PPs, cryptopatches or ILF (Malmstrom et al 2001, Krajina et al 2003). The homing process into the PPs is functionally dependent on the expression of CCR-6, since CCR6$^{-/-}$ mice have reduced numbers of DCs in the SED (Cook et al 2000). Furthermore, CCL9 is secreted by the FAE and mediates the homing of CD11b$^+$ DCs into this region (Zhao et al 2003). After antigen encounter and maturation, intestinal DCs express CCR7 and travel towards the draining lymph node or within the mucosal lymphoid tissue (Shreedhar et al 2003). Tissue factors such as the presence of interleukin (IL)10 can modulate CCR7 expression and therefore influence the homing properties of DCs (Takayama et al 2001). The expression of CCR7 by DCs mediates entry from the lymph into the lymph node via interaction with CCL19 and CCL21 and enables their encounter with naïve T cells.

Naïve T cells home preferentially to the secondary lymphoid organs. High endothelial venules in these organs mediate the entry of lymphocytes via expression of CCL19 and CCL21 (Campbell & Butcher 2002). CCL21 is expressed by endothelial cells of the endothelial venules whereas CCL19 is expressed within T cell and B cell zones and becomes transcytosed to the luminal surface of the endothelium (Campbell & Butcher 2002, Baekkevold et al 2001).

After T cells encounter antigen via interaction of their TCR with MHC–peptide complexes, naïve T cells become activated and primed. Recent studies have shown that T cell activation in the MLNs (mesenteric lymph node) also involves imprinting of gut tropism (Mora et al 2003, Johansson-Lindbom et al 2003, Campbell & Butcher 2002). Thus, CD8$^+$ T cells acquire the gut specific integrin $\alpha 4\beta 7$ and CCR9 expression as a consequence of interactions with DCs (Johansson-Lindbom et al 2003). Associated with the down-regulation of L-selectin, T cells leave the lymphoid tissue and travel to effector sites in the intestinal LP (Johansson-Lindbom et al 2003). Similar mechanisms apply for CD4$^+$ T cells

(Campbell & Butcher 2002). In the lamina propria, T cells act as effector cells but can re-enter the circulation via the lymph (Gowans & Knight 1964, Kunkel et al 2003).

Influence of inflammatory stimuli on the homing of intestinal DCs and $CD4^+$ T cells

Chronic intestinal inflammation in patients with Crohn's disease and ulcerative colitis (O'Neil & Steidler 2003, Banks et al 2003, Yang et al) as well as in mouse models of inflammatory bowel disease (IBD) (Scheerens et al 2001) has been shown to be associated with the up-regulation of a variety of different integrins and integrin ligands as well as chemokines and chemokine receptors. Although the possible role of most of these molecules can be explained by the need to attract increased numbers of leukocytes into the tissue, the functional role during intestinal inflammation still needs to be established for most of these molecules.

The interaction of the $\alpha 4\beta 7$ integrin with its tissue adhesion molecule MAdCAM has been studied in different model systems. Blocking of $\alpha 4\beta 7$ and/or MAdCAM by administration of monoclonal antibodies against these molecules was able to prevent the development of colitis in the T cell transfer model (Picarella et al 1997) as well as the spontaneously occurring colitis in cotton-top tamarin (Hesterberg et al 1996). Somewhat contrary to this, $\beta 7$-deficient $CD4^+CD45Rb^{high}$ cells were still able to induce intestinal inflammation in the T cell transfer model and colitis developed in $IL2^{-/-}\beta 7^{-/-}$ double knockout mice suggesting that the absence of the $\beta 7$ subunit alone is not sufficient to stop the entry of T cells into the mucosa (Sydora et al 2002). These findings identify homing associated integrins as potential targets for therapeutic intervention in patients with IBD. Initial human trials have shown that blockade of $\alpha 4$-integins with the monoclonal antibody Natazulimab can reduce the severity of Crohn's disease and ulcerative colitis (Ghosh et al 2003, Gordon et al 2001, 2002).

In a rat model of TNBS-induced colitis, CCL5 (RANTES) has been shown to be functionally involved in the development of colitis, since CCL5 and its receptors CCR1 and CCR5 were not only up-regulated, but the RANTES antagonist Met-RANTES reduced the intestinal pathology (Ajuebor et al 2001). In a model of DSS-induced colitis, lack of CCR5 has been shown to promote a Th1 to Th2 switch, which reduces the severity of colitis (Andres et al 2000).

Recently, it has been found that a lack of CCR9 on potentially pathogenic $CD4^+CD45Rb^{high}$ T cells prevents the development of colitis in the T cell transfer system of colitis (Manlongat 2003).

The available data underscore the role of homing associated molecules in the development of colitis and highlight the possibility of selectively targeting pathogenic leukocyte populations via their homing mechanisms.

Role of $CD4^+CD25^+$ regulatory T cells in the control of intestinal inflammation

$CD4^+CD25^+$ T_{reg} cells have been shown to prevent the development of intestinal immunopathology (Maloy & Powrie 2001, Sakaguchi et al 2001, Shevach 2002). We investigated the therapeutic potential of T_{reg} cells in mice with established colitis induced by $CD4^+CD45RB^{high}$ T cell transfer (Mottet et al 2003).

The adoptive transfer of $CD4^+CD25^+$ T cells into colitic mice led to recovery from wasting disease indicated by a reduced colonic epithelial cell hyperplasia, a reduction of the LP $CD4^+$ T cell infiltrate and the reappearance of goblet cells. Using a congenic marker to distinguish the pathogenic $CD4^+CD45RB^{high}$ progeny from the $CD4^+CD25^+$ progeny we could show that two weeks after injection of the $CD4^+CD25^+$ T_{reg} cells, the progeny were present in low frequency (1–5% of total $CD4^+$ T cells) in spleen, MLNs and the inflamed colon. Ten weeks after the secondary transfer of T_{reg} cells, the frequency of T_{reg} cells increased to 41% in the MLN and 18% in the LP. $CD4^+CD25^+$ T cells may therefore act not only in the draining MLN but also in the inflamed colon. T_{reg} cells were found *in situ* at the interface of APC and effector T cells. These findings are in line with the hypothesis, that the T_{reg} cell expansion and activity is driven by the inflammation that they regulate (Maloy & Powrie 2001) and that APC–T_{reg} interactions are involved in the regulation of pathogenic T cell responses.

The finding of T_{reg} cells within lymphoid organs as well as in the inflamed colon prompts the question about the mechanisms of their homing. The ability to home towards the lymph node and to the inflamed tissue could be mediated by lymphoid tissue- and inflammation-specific chemokine receptors. Specific migratory activity of T_{reg} cells is suggested by the finding that B cells and professional APCs can recruit $CCR5^+CD4^+CD25^+$ T_{reg} cells via expression of CCL4 (Bystry et al 2001). In addition, $CD4^+CD25^+CD62L^+$ splenocytes with regulatory capacity *in vivo* expressed CCR7 at high levels and migrated *in vitro* towards the lymphoid tissue chemokines CCL21 (SLC) and CCL19 (ELC, macrophage-inflammatory protein 3β) but not towards the inflammatory CXCR3 chemokine ligands CXCL9 (MIG) and CXCL10 (IP10) or towards the CCR4 ligands CCL22 (macrophage derived chemokine, MDC) and CCL17 (thymus and activation-regulated chemokine, TARC) (Szanya et al 2002). In contrast to this, human $CD4^+CD25^+$ cells expressed CCR4 and CCR8 and responded to the CCR4 ligands CCL22 and CCL17, and to the CCR8 ligand CCL1 (I-309) (Iellem et al 2001). Furthermore, CCL2 (monocyte chemoattractant protein 1, MCP-1) attracted murine IL10-producing $CD4^+CD25^+$ T cells *in vitro* (Goulvestre et al 2002). Although still difficult to interpret, since they are derived from different cell populations and model conditions, the current data increasingly indicate that $CD4^+CD25^+$ T cells home towards both lymphoid- and inflammation-associated chemokines.

Differential activation states of the $CD4^+CD25^+$ T cell population and the heterogeneity within the $CD4^+CD25^+$ population may influence experimental results and explain partially contrasting results of the current studies.

Outlook

The further understanding of the molecular mechanisms that are involved in the homing of intestinal immune cells to the mucosal effector site and the MALT may help to develop therapeutic strategies for chronic intestinal inflammation such as ulcerative colitis or Crohn's disease. The ability of regulatory T cell populations to impede pathogenic T cell responses by active regulation might be of therapeutic potential for the treatment of chronic inflammatory disease.

Acknowledgements

H.U. was supported by the Deutsche Forschungsgemeinschaft and the European Society of Clinical Microbiology and Infectious Diseases. C.M. received funding by the Swiss National Science Foundation, the Roche Research Foundation, and the Novartis Foundation. F.P. is supported by the Wellcome Trust. We would like to thank Kevin Maloy and Janine Coombes for critical comments in preparation of this manuscript and the oral presentation at the Novartis Foundation symposium.

References

Ajuebor MN, Hogaboam CM, Kunkel SL, Proudfoot AE, Wallace JL 2001 The chemokine RANTES is a crucial mediator of the progression from acute to chronic colitis in the rat. J Immunol 166:552–558

Andres PG, Beck PL, Mizoguchi E et al 2000 Mice with a selective deletion of the CC chemokine receptors 5 or 2 are protected from dextran sodium sulfate-mediated colitis: lack of CC chemokine receptor 5 expression results in a NK1.1+ lymphocyte-associated Th2-type immune response in the intestine. J Immunol 164:6303–6312

Baekkevold ES, Yamanaka T, Palframan RT et al 2001 The CCR7 ligand elc (CCL19) is transcytosed in high endothelial venules and mediates T cell recruitment. J Exp Med 193:1105–1112

Banks C, Bateman A, Payne R, Johnson P, Sheron N 2003 Chemokine expression in IBD. Mucosal chemokine expression is unselectively increased in both ulcerative colitis and Crohn's disease. J Pathol Bacteriol 199:28–35

Becker C, Wirtz S, Blessing M et al 2003 Constitutive p40 promoter activation and IL-23 production in the terminal ileum mediated by dendritic cells. J Clin Invest 112:693–706

Butcher EC, Picker LJ 1996 Lymphocyte homing and homeostasis. Science 272:60–66

Bystry RS, Aluvihare V, Welch KA, Kallikourdis M, Betz AG 2001 B cells and professional APCs recruit regulatory T cells via CCL4. Nat Immunol 2:1126–1132

Campbell DJ, Butcher EC 2002 Rapid acquisition of tissue-specific homing phenotypes by CD4(+) T cells activated in cutaneous or mucosal lymphoid tissues. J Exp Med 195:135–141

Campbell DJ, Kim CH, Butcher EC 2003 Chemokines in the systemic organization of immunity. Immunol Rev 195:58–71

Cebra JJ 1999 Influences of microbiota on intestinal immune system development. Am J Clin Nutr 69:1046S–1051S

Cook DN, Prosser DM, Forster R et al 2000 CCR6 mediates dendritic cell localization, lymphocyte homeostasis, and immune responses in mucosal tissue. Immunity 12:495–503

Gao Z, Metz WA 2003 Unraveling the chemistry of chemokine receptor ligands. Chem Rev 103:3733–3752

Ghosh S, Goldin E, Gordon FH et al 2003 Natalizumab for active Crohn's disease. N Engl J Med 348:24–32

Gordon FH, Hamilton MI, Donoghue S et al 2002 A pilot study of treatment of active ulcerative colitis with natalizumab, a humanized monoclonal antibody to alpha-4 integrin. Aliment Pharmacol Ther 16:699–705

Gordon FH, Lai CW, Hamilton MI et al 2001 A randomized placebo-controlled trial of a humanized monoclonal antibody to alpha4 integrin in active Crohn's disease. Gastroenterology 121:268–274

Goulvestre C, Batteux F, Charreire J 2002 Chemokines modulate experimental autoimmune thyroiditis through attraction of autoreactive or regulatory T cells. Eur J Immunol 32: 3435–3442

Gowans JL, Knight EJ 1964 The route of re-circulation of lymphocytes in the Rat. Proc R Soc Lond B Biol Sci 159:257–282

Hesterberg PE, Winsor-Hines D, Briskin MJ et al 1996 Rapid resolution of chronic colitis in the cotton-top tamarin with an antibody to a gut-homing integrin alpha 4 beta 7. Gastroenterology 111:1373–1380

Iellem A, Mariani M, Lang R et al 2001 Unique chemotactic response profile and specific expression of chemokine receptors CCR4 and CCR8 by CD4(+)CD25(+) regulatory T cells. J Exp Med 194:847–853

Iwasaki A, Kelsall BL 2000 Localization of distinct Peyer's patch dendritic cell subsets and their recruitment by chemokines macrophage inflammatory protein (MIP)-3alpha, MIP-3beta, and secondary lymphoid organ chemokine. J Exp Med 191:1381–1394

Iwasaki A, Kelsall BL 2001 Unique functions of CD11b+, CD8 alpha+, and double-negative Peyer's patch dendritic cells. J Immunol 166:4884–4890

Johansson-Lindbom B, Svensson M, Wurbel MA et al 2003 Selective generation of gut tropic T cells in gut-associated lymphoid tissue (GALT): requirement for GALT dendritic cells and adjuvant. J Exp Med 198:963–969

Kanamori Y, Ishimaru K, Nanno M et al 1996 Identification of novel lymphoid tissues in murine intestinal mucosa where clusters of c-kit+ IL-7R+ Thy1+ lympho-hemopoietic progenitors develop. J Exp Med 184:1449–1459

Kraehenbuhl JP, Neutra MR 2000 Epithelial M cells: differentiation and function. Annu Rev Cell Dev Biol 16:301–332

Krajina T, Leithauser F, Moller P, Trobonjaca Z, Reimann J 2003 Colonic lamina propria dendritic cells in mice with CD4+ T cell-induced colitis. Eur J Immunol 33:1073–1083

Kunkel EJ, Campbell DJ, Butcher EC 2003 Chemokines in lymphocyte trafficking and intestinal immunity. Microcirculation 10:313–323

Malmstrom V, Shipton D, Singh B et al 2001 CD134L expression on dendritic cells in the mesenteric lymph nodes drives colitis in T cell-restored SCID mice. J Immunol 166:6972–6981

Maloy KJ, Powrie F 2001 Regulatory T cells in the control of immune pathology. Nat Immunol 2:816–822

Manlongat NTV, Mak S, Manning S, Picarella D 2003 Exploring the role of gastrointestinal homing receptors in murine models of inflammatory bowel disease. New York Academy of Sciences Conference. Oral tolerance: mechanisms and applications (abstr 19)

Mebius RE 2003 Organogenesis of lymphoid tissues. Nat Rev Immunol 3:292–303

Mora JR, Bono MR, Manjunath N et al 2003 Selective imprinting of gut-homing T cells by Peyer's patch dendritic cells. Nature 424:88–93

Mottet C, Uhlig HH, Powrie F 2003 Cutting edge: cure of colitis by CD4+CD25+ regulatory T cells. J Immunol 170:3939–3943

Mowat AM 2003 Anatomical basis of tolerance and immunity to intestinal antigens. Nat Rev Immunol 3:331–341

Neutra MR, Pringault E, Kraehenbuhl JP 1996 Antigen sampling across epithelial barriers and induction of mucosal immune responses. Annu Rev Immunol 14:275–300

O'Neil D, Steidler L 2003 Cytokines, chemokines and growth factors in the pathogenesis and treatment of inflammatory bowel disease. Adv Exp Med Biol 252–285

Picarella D, Hurlbut P, Rottman J et al 1997 Monoclonal antibodies specific for beta 7 integrin and mucosal addressin cell adhesion molecule-1 (MAdCAM-1) reduce inflammation in the colon of scid mice reconstituted with CD45RBhigh CD4+ T cells. J Immunol 158:2099–2106

Sakaguchi S, Sakaguchi N, Shimizu J et al 2001 Immunologic tolerance maintained by CD25+ CD4+ regulatory T cells: their common role in controlling autoimmunity, tumor immunity, and transplantation tolerance. Immunol Rev 182:18–32

Scheerens H, Hessel E, de Waal-Malefyt R, Leach MW, Rennick D 2001 Characterization of chemokines and chemokine receptors in two murine models of inflammatory bowel disease: IL-10−/− mice and Rag-2−/− mice reconstituted with CD4+CD45RBhigh T cells. Eur J Immunol 31:1465–1474

Shevach EM 2002 CD4+ CD25+ suppressor T cells: more questions than answers. Nat Rev Immunol 2:389–400

Shreedhar VK, Kelsall BL, Neutra MR 2003 Cholera toxin induces migration of dendritic cells from the subepithelial dome region to T- and B-cell areas of Peyer's patches. Infect Immun 71:504–509

Sydora BC, Wagner N, Lohler J et al 2002 beta7 Integrin expression is not required for the localization of T cells to the intestine and colitis pathogenesis. Clin Exp Immunol 129:35–42

Szanya V, Ermann J, Taylor C, Holness C, Fathman CG 2002 The subpopulation of CD4+CD25+ splenocytes that delays adoptive transfer of diabetes expresses L-selectin and high levels of CCR7. J Immunol 169:2461–2465

Takayama T, Morelli AE, Onai N et al 2001 Mammalian and viral IL-10 enhance C-C chemokine receptor 5 but down-regulate C-C chemokine receptor 7 expression by myeloid dendritic cells: impact on chemotactic responses and in vivo homing ability. J Immunol 166:7136–7143

Yamanaka T, Helgeland L, Farstad IN et al 2003 Microbial colonization drives lymphocyte accumulation and differentiation in the follicle-associated epithelium of Peyer's patches. J Immunol 170:816–822

Yang S, Choi M, Kim O et al 2002 The increased expression of an array of C-X-C and C-C chemokines in the colonic mucosa of patients with ulcerative colitis: regulation by corticosteroids. Am J Gastroenterol 97:126–132

Yoshida M, Watanabe T, Usui T et al 2001 CD4 T cells monospecific to ovalbumin produced by Escherichia coli can induce colitis upon transfer to BALB/c and SCID mice. Int Immunol 13:1561–1570

Zhao X, Sato A, Dela Cruz CS et al 2003 CCL9 is secreted by the follicle-associated epithelium and recruits dome region Peyer's patch CD11b+ dendritic cells. J Immunol 171:2797–2803

DISCUSSION

Moore: I have a question addressed to both Holm Uhlig and Sander Van Deventer. Do you need the cells or can you get away with delivering TGFβ and IL10?

Van Deventer: The only thing that works is IL10 and cells. That's all we can say so far. IL10 without cells doesn't work.

Moore: So you need T_{reg} cells.

Van Deventer: Even with probiotics that produce IL10, we think it is *in situ* maturation — it is not the IL10 from the bacteria, but the IL10 from the T_{regs} that are induced by the disease. I don't know whether it is the secreted IL10 or whether it is IL10 in the contact–contact situation, but most of our data suggest the latter. Maria Grazia Roncarolo also thinks this.

Parkos: I noticed that you said that cryptopatches aren't present in the human mucosa.

Uhlig: I am not aware of any studies that have shown the cryptopatch phenotype in humans. I know that a lot of people are working on this.

Parkos: Does anyone know what the epithelial receptor is for the $CD11c^+$ DCs? Is it CD11c that binds to something on the epithelium?

Uhlig: CCL20 is produced and this is probably in close contact with CCR6. This fits with data from Brian Kelsall who has shown that the sub-epithelial DCs are CCR6 positive (Iwasaki & Kelsall 2000).

Parkos: Has anyone done adhesion-based assays with these DCs?

Uhlig: I think that Brian Kelsall has done this (Iwasaki & Kelsall 2000). The CCR6 knockout mouse seems to be the best indication that CCR6–CCL20 interactions play a role. CCR6 knockouts have a depletion of $CD11b^+CD11c^+$ in the subepithelial dome area (Cook et al 2000).

Parkos: Is there a permeability defect in the gut of these mice?

Powrie: When they have IBD there is. Do you mean is there altered permeability in the SCID mice?

Parkos: Yes.

Powrie: I am not aware of any data that have shown that. I think people are looking at tight junctions and things like that.

Jewell: Holm Uhlig, were you implying that the homing mechanism of T_{reg} cells is through CCR6 and so on, and that they don't use the $\alpha 4\beta 7$ MAdCAM mechanism?

Uhlig: I don't know. I would like to keep this open. I just wanted to make the point that there are several studies with controversial results, and this might be because there are different cells that people regard as regulatory cells. They may have different activation states, and there are human versus mouse differences.

Kelleher: I have a question concerning TGFβ in this system. One of the things that TGFβ does traditionally is to increase expression of CD103. Do the TGFβ producing cells alter the expression of CD103 on your $CD11c^+CD103^-$ DCs?

Uhlig: I was looking at the subepithelial and epithelial dome region, an area described in other studies as being very rich in TGFβ. I hoped that intraepithelial DCs would be $CD103^+$, and then it would have been nice to say that E-cadherin, the ligand for CD103, was present in the epithelium. Unfortunately the intraepithelial DCs were CD103 negative. This finding at least does not support a role for CD103 in the homing of DCs to the epithelium.

Kelleher: The other thing about TGFβ is that it comes as a precursor with latency-associated peptide (LAP). LAP has RGD binding sequences, which could potentially complex with α4. In other systems they do bind to $\alpha\beta$. Is this a potential mechanism for the regulatory effect in the gastrointestinal tract?

Powrie: It could be. Do you know whether TGFβ has been shown to induce CD103 on DCs?

Kelleher: No. It certainly does on epithelial cells but I have not seen data relating to effects on DCs.

Mahida: When you were following up the T_{regs} in your model, did you get a feel for where they were going first? Are they going to the LP first and then going to the mesenteric lymph nodes, or are they going to these sites at the same time?

Uhlig: We were looking at two weeks after the T_{reg} cell transfer, and at that time these cells were extremely rare. At this relatively early time point these cells are present in both, secondary lymphoid tissue and inflamed tissue. I have no idea where they go first. At least when they are present at low frequencies, they are present in both the colon and mesenteric nodes.

Powrie: We need to do more detailed kinetic analysis to address whether they go to the node and then to the effector site.

Mahida: That has always been the presumption.

Powrie: I don't think our studies have really addressed this.

Mahida: If I remember correctly the cryptopatch work that came out in *Science* (Saito et al 1998) showed that they can differentiate into $CD4^+$ and $CD8^+$ cells. Do you have any views regarding their contribution?

Uhlig: Kanamori and coworkers have shown that SCIDs and RAGs have CD11c-rich clusters (Kanamori et al 1996). The principle structure is there: they just don't contain mature T cells. The accumulation of $CD11c^+$ cells and very low numbers of $CD4^+$ cells are present in SCIDs. The argument that cryptopatches may play a functional role for T cell responses comes from the nude mice that do not have a functional thymus. In these mice cryptopatches and a reasonable proportion of intra-epithelial lymphocytes are present, arguing for thymus-independent T cell development.

Ghosh: You touched on the DCs sampling of luminal antigens through epithelial tight junctions. Is the T cell outcome of such a sampling different from Peyer's patch sampling?

Uhlig: So far we have looked under non-inflammatory conditions. The T cell response against luminal derived antigens is marginal. We wanted to look in an antigen-specific system that mimics a response against commensal bacteria. We see that there are T cells accumulating at low numbers within colonic lymphoid clusters. Since we don't see strong T cell responses at all under non-inflammatory conditions, I can't comment on whether there would be a relevant difference between DC subpopulations.

Van Deventer: Can you explain to me why there are so many $CCR9^+$ T cells in the colon? The ligand is not expressed there. Are these all memory cells or are there also many $CCR9^+$ cells in the colon?

Uhlig: There is work by Eric Kunkel (Kunkel et al 2000) where they looked for CCR9 expression in the colon and small intestine LP. They found 25% of colonic $CD4^+$ T cells to be positive for CCR9. In the small intestine almost all the LP T cells are positive. I am not aware of anyone having analysed the colonic T cell subpopulations for CCR9.

Van Deventer: Do you think T cells can be primed in the lamina propria by DCs? Classically they need to go to the lymph nodes.

Mahida: We have done some work showing that naïve T cells can be stimulated by lamina propria DCs.

Van Deventer: How important is that? I ask because the clinicians here know that, for example, if you give anti-TNF and clear the ulcers, when the ulcers recur (which they usually do) they are the same ulcers. What kind of resident cell is present that is causing that?

Powrie: Given these leukocyte clusters with organized structures in the LP it is possible that T cells could be primed there. Joerg Reimann's data using the T cell transfer model show proliferation very early on in clusters in the colonic mucosa. Anti-TNF could be having an important role in disrupting these types of structures.

Van Deventer: Where are the T_{regs} made? In the LP or in the lymph nodes?

Powrie: Naturally arising T_{regs} differentiate in the thymus but it is possible that cells like these or Tr1 cells may also develop in effector sites. There is not much information on this.

Jewell: Sander Van Deventer, do you want to answer your own question?

Van Deventer: I think it is an important question that isn't frequently addressed. I think that most of the priming and regulation might be in the LP, rather than in the lymph node.

Sartor: But what is the evidence in a non-lymphoid aggregate that there is a naïve T cell that isn't yet stimulated? I can understand local clonal expansion and stimulation, but is there evidence for a truly naïve T cell that hasn't yet undergone a T cell receptor expansion?

Van Deventer: I think they are there.

Mahida: Phenotypically, if you isolate T cells there are some naïve ones present, although the majority are memory T cells.

Sartor: By isolation, how can you exclude the small aggregates? You could cut out Peyer's patches.

Van Deventer: In situ they are present.

Mahida: How frequent are aggregates in human colon?

Uhlig: There are 20 000 or so.

Mahida: I have a comment relating to CCR6. It is a ligand for HBD1 and HBD2 in immature DCs, but I didn't see any reference to intervention where you lost CCR6 function.

Powrie: I don't think that there is any work on this. It is notable that anti-CCR9 does present colitis in this model. This doesn't really fit with where the ligand is, as Sander Van Deventer pointed out earlier. This may say something about the site of priming.

Van Deventer: I was thinking about that strange experiment, and thought that they must be primed in the small bowel and then they recirculate through $\alpha 4\beta 7$, or something like that. The effector ones are CCR9-positive, but it is not the CCR9 that brings them there. But then it is still difficult to understand why the CCR9 blockade is effective.

Powrie: There is something we don't understand here.

References

Cook DN, Prosser DM, Forster R et al 2000 CCR6 mediates dendritic cell localization, lymphocyte homeostasis, and immune responses in mucosal tissue. Immunity 12:495–503

Iwasaki A, Kelsall BL 2000 Localization of distinct Peyer's patch dendritic cell subsets and their recruitment by chemokines macrophage inflammatory protein (MIP)-3α, MIP-3β, and secondary lymphoid organ chemokine. J Exp Med 191:1381–1394

Kanamori Y, Ishimaru K, Nanno M et al 1996 Identification of novel lymphoid tissues in murine intestinal mucosa where clusters of c-kit$^+$ IL-7R$^+$ Thy1$^+$ lympho-hemopoietic progenitors develop. J Exp Med 184:1449–1459

Kunkel EJ, Campbell JJ, Haraldsen G et al 2000 Lymphocyte CC chemokine receptor 9 and epithelial thymus-expressed chemokine (TECK) expression distinguish the small intestinal immune compartment: Epithelial expression of tissue-specific chemokines as an organizing principle in regional immunity. J Exp Med 192:761–768

Saito H, Kanamori Y, Takemori T et al 1998 Generation of intestinal T cells from progenitors residing in gut cryptopatches. Science 280:275–278

Anti-TNF therapy in Crohn's disease

S. Ghosh

Gastroenterology Section, Imperial College London, Hammersmith Hospital, Ducane Road, London W12 0NN, UK

Abstract. Anti-tumour necrosis factor (TNF) strategies, the most studied of biological therapies, include chimeric monoclonal (infliximab), humanized monoclonal (CDP571 and the PEGylated CDP870) and fully human monoclonal (adalimumab) antibodies, p75 fusion protein (etanercept), p55 soluble receptor (onercept) and small molecules such as MAPkinase inhibitors. The principal use of infliximab is in treating active Crohn's disease patients not responding to or intolerant of conventional therapies. Infliximab is steroid sparing. The development of antibodies against infliximab is associated with an increased risk of infusion reactions and a reduced duration of response to treatment, and concomitant immunosuppressive therapy reduces the immunogenic response. The demonstration of the efficacy of maintenance therapy every 8 weeks with infliximab in the randomised, controlled, ACCENT I trial opened up the strategy for regular maintenance. In patients who have failed therapy with cortocosteroids and immunosuppressive therapy and are poor surgical candidates, and patients with fistulizing disease, where infliximab therapy is chosen, regular maintenance therapy with infliximab is likely to be required. On the other hand, patients with severely active, steroid-refractory disease in whom immunosuppressive therapy and infliximab are initiated together, may respond adequately and be continued on long-term immunosuppressive therapy alone. In ulcerative colitis the role of infliximab remains uncertain.

2004 Inflammatory bowel disease — crossroads of microbes, epithelium and immune systems. Wiley, Chichester (Novartis Foundation Symposium 263) p 193–210

Anti-tumour necrosis factor (TNF) therapy has been most studied amongst biological therapies in inflammatory bowel disease (IBD). Anti-TNF strategies include chimeric monoclonal antibody (infliximab), humanized monoclonal antibody (CDP 571 and the PEGylated CDP870), fully human monoclonal antibody (adalimumab), p75 fusion protein (etanercept), p55 soluble receptor (onercept) and small molecules such as MAP kinase inhibitors and thalidomide congeners. Such therapies fulfil a felt need in current IBD management armamentarium (Table 1).

The role of TNF in Crohn's disease

Crohn's disease (CD) is currently best understood as an activation of lamina propria macrophages and dendritic cells driven by intestinal luminal bacterial

TABLE 1 Why do we need biological therapy?

• Corticosteroids	• Immunosuppressives
– Resistance	– Slow onset
– Dependence	– Non response
– No maintenance	– Serious side effects
– Lack of mucosal healing	– Intolerance
– Side effects	

antigens leading to a T helper 1 (Th1) lymphocyte proliferation. TNF has been demonstrated to be elevated in the blood, stool and intestinal tissues of patients with CD (Murch et al 1991, 1993, Braegger et al 1992). Our current understanding of the immunology of CD is illustrated in Figs 1a and 1b.

Mechanism of action of anti-TNF strategies

Apart from neutralizing soluble cytokines, monoclonal antibodies may deplete immune cells by a number of mechanisms that include: (a) antibody-dependent cell-mediated cytotoxicity (ADCC); (b) complement-dependent cytotoxicity (CDC); (c) apoptosis of immune cells, lymphocytes or monocytes.

Infliximab binds to both soluble and transmembrane TNF, and also results in lysis of immune cells expressing TNF via CDC and ADCC (Papadakis & Targan 2000, Scallon et al 1995). Considerable interest has focused on *in vitro* and *in vivo* demonstration of apoptosis of lymphocytes and monocytes in the intestinal lamina propria rapidly after exposure to infliximab (Lugering et al 2001, ten Hove et al 2002). Etanercept neutralizes soluble TNF but does not bind to transmembrane TNF. CDP571 neutralizes both soluble and transmembrane TNF. CNI-1493 inhibits MAP kinase, especially c-Jun N-terminal kinase (JNK) signalling pathways that leads to activation of the TNF gene. Thalidomide congeners inhibit TNF synthesis by inhibiting interleukin (IL)12, which promotes Th1 proliferation, as well as degrading TNF mRNA in macrophages. Considerable controversy exists about the exact mechanism of action of anti-TNF therapies in different chronic inflammatory conditions.

Chimeric monoclonal antibody to TNF (infliximab)

Infliximab is a chimeric IgG1 monoclonal antibody that is approximately 75% human and 25% murine. Infliximab has been shown to be effective in the treatment of moderate to severe active and fistulizing CD. The principal use of infliximab is in treating patients not responding to conventional therapies or

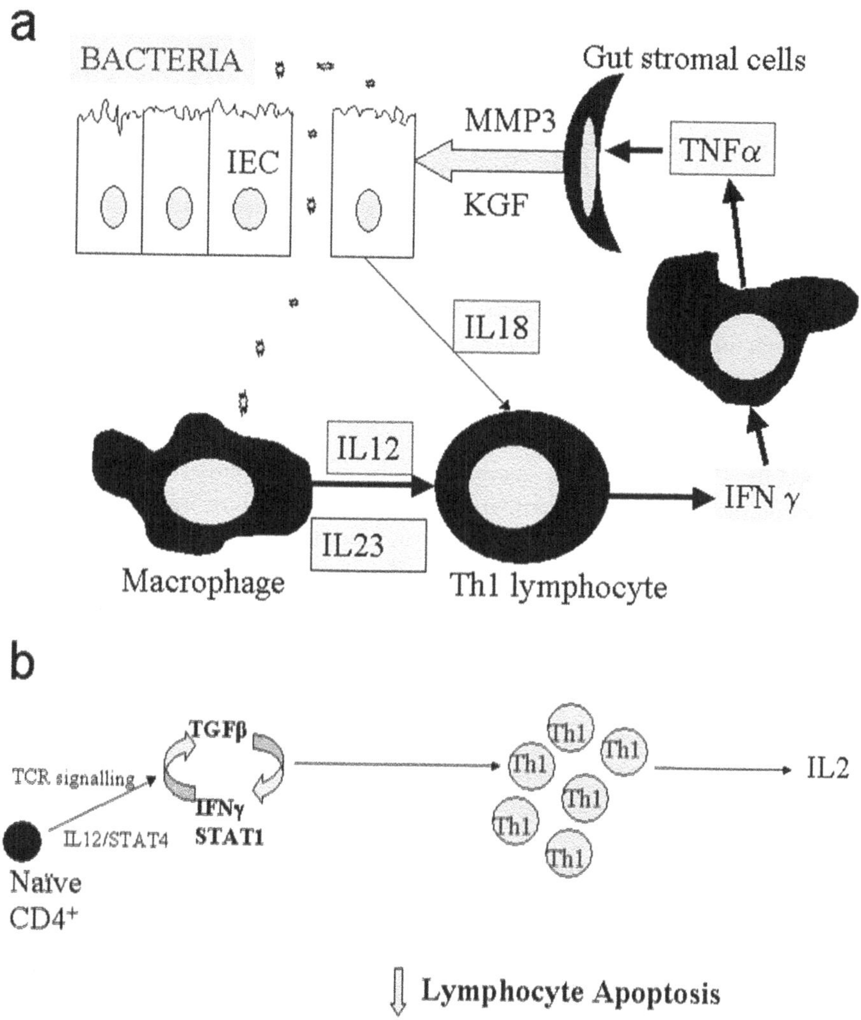

FIG. 1. (a) Luminal bacterial antigen-driven activation of mucosal immune cells and production of pro-inflammatory cytokines including TNF. (b) Crohn's disease is characterized by a proliferation of Th1 lymphocytes in the intestinal lamina propria and defective apoptosis of lymphocytes leading to persistence of activated lymphocytes.

developing unacceptable side effects. A single infusion of infliximab resulted in an impressive short-term response in a 12-week multicenter, double-blind placebo-controlled trial of infliximab (Targan et al 1997). A dose of 5 mg/kg was most effective, with 81% of patients (vs. 17% receiving placebo) showing a clinical response (decrease in CDAI ≥ 70 points) and 48% (vs. 4% receiving placebo) entering a clinical remission (CDAI < 150) at 4 weeks (Fig. 2). Longer-term therapy, using 10 mg/kg given every 8 weeks in four infusions to initial responders to infliximab, was effective in maintaining clinical benefit throughout the re-treatment period and 8 weeks after the last infusion (53% at week 44), but the placebo group had only 20% of patients in remission at week 44 (Rutgeerts et al 1999). Reports of the usefulness of infliximab in clinical practice outside trials confirm its efficacy and safety. Infliximab is steroid sparing, and in the Mayo clinic experience steroids could be discontinued in 73% of patients. Approximately one-third of patients administered a second infusion of infliximab after not responding to a first dose may improve clinically. Such episodic treatment does not factor in the higher immunogenicity of such treatment and potential loss of efficacy requiring higher or more frequent doses, or higher hospital or surgical costs, as a consequence of loss of efficacy. In a review of infliximab use in Chicago, annual incidence of surgeries were noted to have declined 38%, endoscopies declined by 43%, radiological examinations by 12% and outpatient GI visits by 20%—all significantly less compared to the year preceding infliximab use. In patients with fistulizing CD, there was a significant decrease in hospitalizations (by 59%), in GI surgeries (by 59%) and radiological examinations by 40%. Such reports are very encouraging, as a decrease in healthcare resources raises the potential of overall cost savings in CD patients receiving infliximab.

The development of antibodies against infliximab is associated with an increased risk of infusion reactions and a reduced duration of response to treatment. Concomitant immunosuppressive therapy reduces the magnitude of the immunogenic response. Both concurrent immunosuppressive therapy and intravenous hydrocortisone significantly reduce antibodies to infliximab (ATI). The demonstration of the efficacy of maintenance therapy every 8 weeks with infliximab in the randomized, controlled, ACCENT I trial on 573 patients provided support for the regular maintenance strategy rather than episodic therapy in CD patients who responded to an initial dose of infliximab (Hanauer et al 2002). In such patients, infliximab therapy, if maintained every 8 weeks for 1 year, resulted in a greater likelihood of remaining in remission at weeks 30 and 54, discontinuing corticosteroids, and maintaining response for a longer period of time with a better quality of life (Fig. 3). The ACCENT I study was pivotal in establishing maintenance treatment with infliximab in CD (Hanauer et al 2002). In patients who have failed therapy with corticosteroids and immunosuppressive therapy and are poor surgical candidates, regular maintenance therapy with infliximab is likely to be required. Patients with fistulizing CD, where infliximab therapy is chosen, are also likely to require maintenance therapy. Both the initial study (Present et al 1999) and the more recent larger re-treatment study (Sands et al 2002) for fistulizing disease showed both a short time and a longer time response with fistula closure if treatment was maintained. Of the

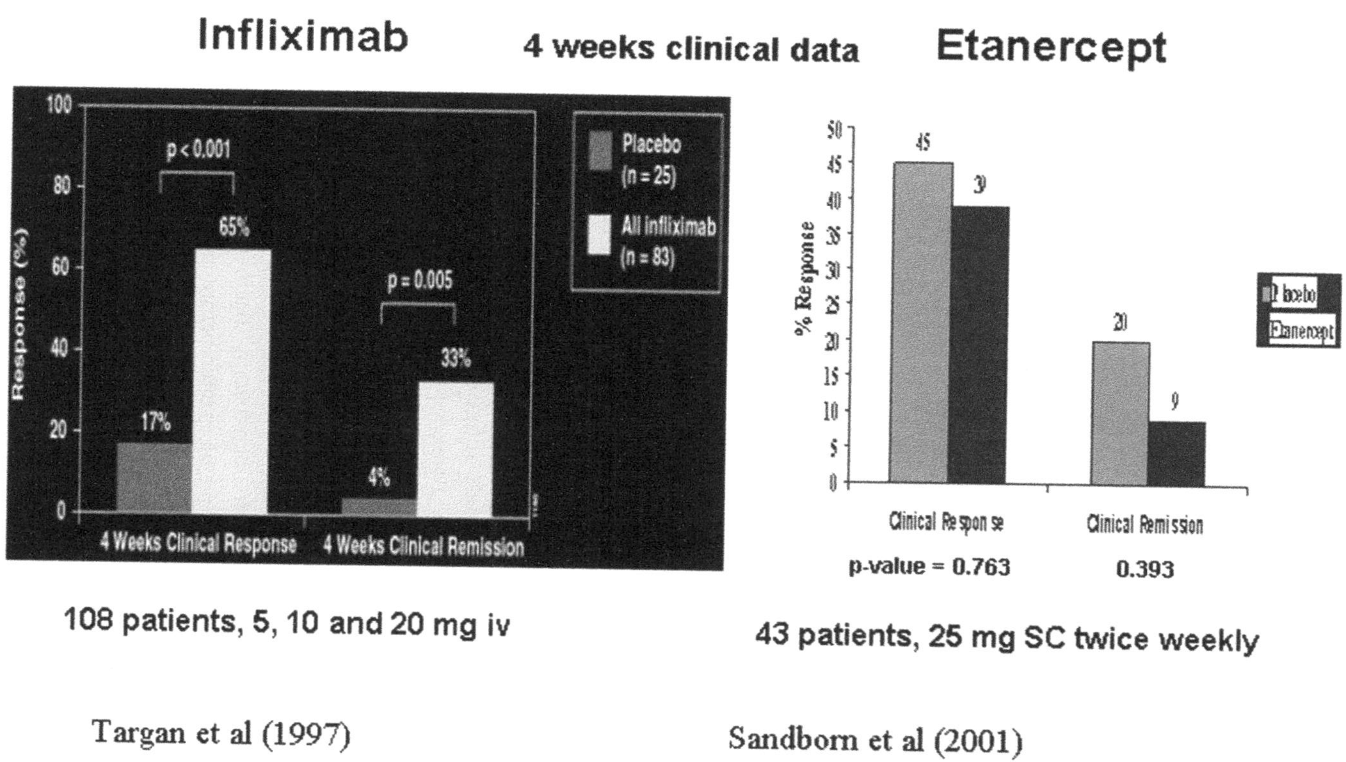

FIG. 2. Short-term efficacy of infliximab in terms of clinical response and remission and contrasting lack of efficacy of etanercept.

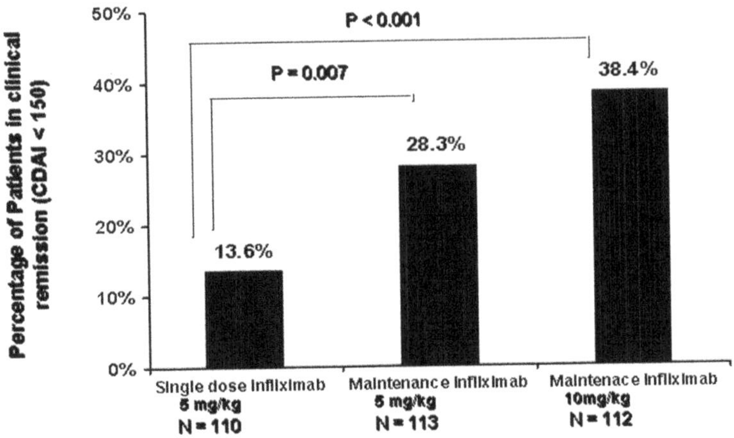

FIG. 3. Summary of ACCENT I maintenance trial of infliximab showing week 54 remission rate amongst week 2 responders (Hanauer et al 2002).

306 patients, 195 (69%) had a fistula response at week 14. The median time to loss of response through week 54 was significantly greater for patients treated with infliximab at 5 mg every 8 weeks (>40 weeks) compared with 14 weeks for placebo-treated patients. CD of the ileoanal pouch may also respond favourably and may close fistulous tracts to adjoining organs (Ricart et al 1999). On the other hand, patients with severely active, steroid-refractory CD disease in whom immunosuppressive therapy and infliximab are initiated together, may respond adequately and be continued on long-term immunosuppressive therapy alone. Subanalysis of the ACCENT I study suggests that scheduled (every 8 weeks) infliximab therapy, particularly the 10 mg/kg group, had better CDAI and IBD quality of life (IBDQ) responses than those in the episodic therapy group. All the 8 week scheduled therapy groups (5 mg/kg and 10 mg/kg) had fewer hospitalizations, higher rates of mucosal healing, and fewer developed antibodies than those in the episodic therapy group (Rutgeerts et al 2004).

Immunogenicity is a significant argument against the episodic use of infliximab (Table 2). In a study of episodic infliximab therapy (Baert et al 2003), the presence of ATI at concentrations of 8 μg/ml or greater before an infusion predicted a shorter duration of response (35 days, as compared with 71 days among patients with concentrations of less than 8 μg/ml; $P < 0.001$) and a higher risk of infusion reactions (relative risk, 2.40; 95% confidence interval, 1.65–3.66; $P < 0.001$). Patients who had infusion reactions had a median duration of clinical response of 38.5 days, as compared with 65 days among patients who did not have an infusion reaction ($P < 0.001$). Concomitant immunosuppressive therapy was predictive of

TABLE 2 Comparison of infliximab regimens

• Episodic	• Maintenance
– Fewer infusions	– Less immunogenic
– Lower cost	– Better mucosal healing
– More immunogenic	– Better QoL
– Poorer QoL	– Less hospitalization (?)
– Less mucosal healing	– Less surgery (?)
– Less predictable	– Higher cost
	– Uncertainty after 1 year

low titres of antibodies against infliximab ($P < 0.001$) and high concentrations of infliximab four weeks after an infusion ($P < 0.001$).

In ulcerative colitis the role of infliximab remains uncertain, and large phase III studies are ongoing. Anecdotal and open-labelled pilot studies had reported encouraging results and some encouraging results have also been obtained with other anti-TNF agents such as CDP571. However, in the only sizeable randomized controlled trial, 43 patients with steroid-refractory ulcerative colitis received infliximab or placebo at weeks 0 and 2 (Probert et al 2003). At week 6, 39% of infliximab treated patients and 30% of placebo treated patients achieved remission (P = not significant), and neither the primary endpoint nor a number of secondary endpoints (improvement in ulcerative colitis symptom score, quality of life or histological improvement) reached statistical significance.

Humanized monoclonal antibody to TNF (CDP571 and CDP870)

CDP571 is a humanized monoclonal IgG4 antibody to TNF with approximately 95% human residues. The 5% murine part of CDP571 is the complementarity determining region (CDR) within the variable region. Initial dose-ranging studies suggested that 10 mg/kg of CDP571 was the most effective dose for short-term clinical response, with a trend towards maintenance of remission efficacy (Sandborn et al 2001a). In a 16-week study (Feagan et al 2000), twice the proportion of patients discontinued steroids on CDP571 (20 mg/kg at baseline followed by 10 mg/kg at week 8) compared with placebo therapy (44% vs. 22%; $P < 0.05$). 396 adult active CD patients (CDAI 220–450) were randomized to receive intravenous CDP571 10 mg/kg ($n = 264$) or placebo ($n = 132$).

CDP 870 is a Fab′ fragment of a humanized anti-TNF monoclonal antibody with added polyethylene glycol to increase the plasma half-life. In a phase II randomized, placebo-controlled study 292 adult patients with active CD (CDAI 220–450) received subcutaneous CDP870 (100, 200 or 400 mg) or placebo. The

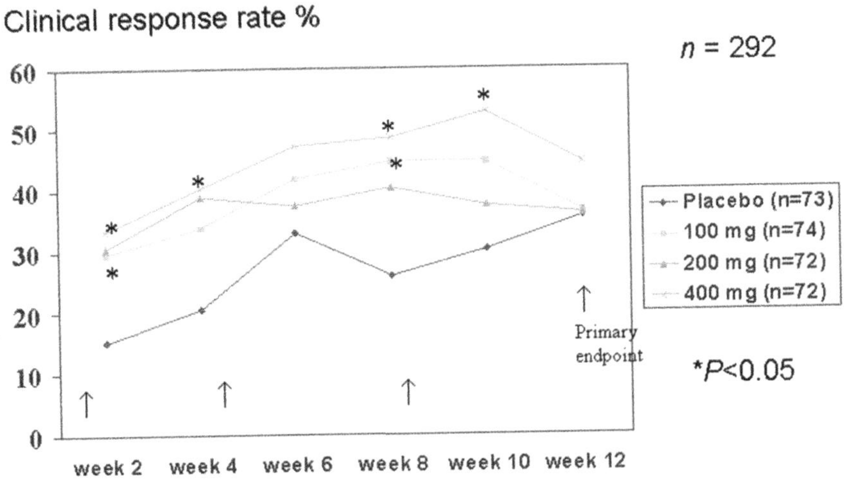

FIG. 4. CDP870 trial establishing 400 mg as the optimum dose administered subcutaneously (Schreiber et al 2003a).

treatment was administered at weeks 0, 4 and 8. Clinical response (decrease in CDAI$\geqslant$100) and remission (CDAI$\leqslant$150) were assessed every 2 weeks up to week 12. Clinical response rates were highest in the CDP870 400 mg treatment group at all time points (Fig. 4). Onset of effect of CDP870 was evident at week 2 (Schreiber et al 2003a). Multivariate analysis identified baseline C-reactive protein (CRP) as predictive of significant response. Only 41% of patients (119/290) had baseline CRP$\geqslant$10 mg/l. In the 'overall intention to treat' analysis CDP870 400 mg was the most effective dose. In the patient subgroup with CRP$\geqslant$10 mg/l, CDP870 400 mg induced a significantly higher clinical response and remission compared to placebo at all time points. Week 12 clinical response was 53.1% (17/32) vs. placebo 17.9% (5/28) (P=0.005). No advantage could be seen with higher doses of CDP870 administered intravenously.

Fully human monoclonal antibody to TNF (Adalimumab)

Adalimumab (D2E7) is a fully human IgG1 monoclonal antibody. Phase II trials are currently ongoing to determine the efficacy of adalimumab for the treatment of active CD.

Soluble TNF receptor strategies (Etanercept and Onercept)

Etanercept. Etanercept is a completely human fusion protein consisting of the human soluble TNF receptor (p75) linked to the Fc portion of an IgG1 antibody.

Etanercept has been shown to be ineffective in the treatment of CD (Sandborn et al 2001b).

Onercept. This is human soluble TNF receptor (p55). Administration of 50 mg three times per week for 12 weeks showed clinical remission (CDAI < 150) in 67% of patients in a small study (Rutgeerts et al 2001). Results from a larger phase II placebo-controlled study are awaited. However soluble TNF receptor strategies may not appear to be effective in CD, unlike rheumatoid arthritis or ankylosing spondylitis.

Peptide molecules

RDP58. This is a decapeptide composed of nine D-amino acids and glycine and has been shown to inhibit the synthesis of TNF and interferon γ (Iyer et al 2002). RDP58 interferes with TNF synthesis at a translational step. Anti-inflammatory effects demonstrated in animal models have now led to clinical trials, which will be reported soon.

Non-protein small molecules

MAP kinase inhibitors. Inflammatory MAPKs play an important role in the pathogenesis of CD and their inhibition may provide an effective therapeutic strategy. An initial small short-term study using intravenous CNI-1493 (a guanylhydrazone) 8 or 25 mg/m^2 in 12 patients with severe CD demonstrated significant reduction in mucosal TNF production and JNK phosphorylation. This was associated with significant clinical benefit and endoscopic improvement (Hommes et al 2002). Inhibition of p38 MAPK however has failed to ameliorate experimental colitis and further studies with p38MAPK and JNK inhibitors are awaited. Toxicity of MAP kinase inhibitors especially hepatoxicity may preclude immediate clinical applications.

Thalidomide congeners. Thalidomide inhibits production of TNF by monocytes. Thalidomide may heal severe refractory oral ulceration caused by CD at a dose of 0.7 mg/kg (Weinstein et al 1999). Open-labelled 12 week trials of thalidomide in refractory CD have reported symptomatic benefit (Ehrenpreis et al 1999, Vasiliauskas et al 1999). In addition to its anti-TNF effect, thalidomide has other potential mechanisms of action including inhibition of neovascularization.

Anti-TNF therapy in extraintestinal manifestations of CD

Both infliximab and etanercept are effective in ankylosing spondylitis. Infliximab results in rapid improvement and a durable response for at least a year (Braun et al

2003). Infliximab was administered at a dose of 5 mg/kg every 6 weeks after induction of remission by 0, 2 and 6 week administrations. Anecdotal reports also demonstrate encouraging efficacy in healing pyoderma gangrenosum.

Prediction of response

Considerable efforts have focused on prediction of response to infliximab. Non-smoking and concurrent immunosuppressive use are associated with a higher rate of response and longer duration of response (Parsi et al 2002). In particular NOD2/CARD15 does not influence response to infliximab in CD (Vermeire et al 2002).

Safety issues with anti-TNF therapy

Serious and opportunistic infections may occur after anti-TNF therapy. Infliximab therapy has been associated with tuberculosis, histoplasmosis, listeriosis and aspergillosis, and similar problems have been reported with the use of etanercept in rheumatoid arthritis. Tuberculosis tends to manifest itself generally within the first six months after initiation of therapy, and represents re-activation of latent tuberculosis. Currently patients are screened by a chest radiograph and tuberculin skin testing prior to initiation of infliximab therapy. Out of 500 consecutive patients suffering from CD treated with infliximab at the Mayo Clinic (Colombel et al 2004), 41 patients (8.2%) had infections attributable to infliximab of which 20 patients (4%) had a serious infection (4 deaths). Antibodies to infliximab (ATI) were reported to be present in 28% of patients in an integrated safety data set including both CD and ulcerative colitis (Schaible 2000). The presence of infliximab in serum interferes with ATI detection. ATI is associated with infusion reactions and loss of efficacy. Concurrent therapy with immunosuppressive drugs or intravenous hydrocortisone reduces ATI formation. Infusion reactions, urticaria, dyspnoea or hypotension, occurring within 2 hours following administration happen in about 17% of patients treated with infliximab compared with 7% treated with placebo (Schaible 2000). Infusion reactions can however be prevented or treated in nearly all patients upon re-treatment (Cheifetz et al 2003). Delayed hypersensitivity reactions characterized by myalgia, arthralgia, fever, skin rash, pruritis, angioedema or dysphagia tend to occur after a prolonged interruption of infliximab therapy. Autoantibody formation such as antinuclear antibody (34%) and anti-double stranded DNA (9%) rarely lead to drug-induced lupus (Schaible 2000). Currently the safety data regarding the risk of malignancy appear to be reassuring despite at least four patients each with CD and rheumatoid arthritis developing non-Hodgkin's lymphoma. These chronic inflammatory diseases, especially rheumatoid arthritis, are associated with an increased baseline

TABLE 3 Mnemonic for pre-infliximab checklist

• S=sepsis • T=TB • O=optic neuritis • I=infusion reaction • C=cancer	STOIC

risk of malignant lymphoma. In the Mayo Clinic series, 3 out of the 500 patients developed malignancies which may be linked to infliximab therapy (Colombel et al 2004). Infliximab may exacerbate demyelinating disorders, and rarely may be associated with a new demyelination disorder (Colombel et al 2004). In the Mayo Clinic series, events leading to death were possibly related to infliximab in 5/500 patients (1%) (Colombel et al 2004).

Pooled safety analysis of CDP571 shows that it is well tolerated, with no reports of tuberculosis, opportunistic infections or lymphoma (Jewell et al 2003). Anti-idiotype antibodies and infusion reactions occur in 5.3% and 12.7% of patients, respectively. In the CDP870 studies, both subcutaneous at weeks 0, 4 and 8 (292 patients: 100 mg [n=74], 200 mg [n=72], 400 mg [n=73], placebo [n=73]) and intravenous (single dose 1.25 mg/kg [n=2], 5 mg/kg [n=26], 10 mg/kg [n=17], 20 mg/kg [n=23] or placebo [n=24]) dosing studies have been analysed for pharmacokinetics and safety. Following s.c. and i.v. administration of CDP870, mean maximal plasma concentrations were dose proportional with a plasma half-life of approximately 2 weeks. There was no increase in opportunistic infections or tuberculosis (Schreiber et al 2003b). There was no malignancy or lupus in the two studies. In the s.c. study, a low incidence of mild to moderate injection site reactions not necessitating withdrawal were reported (2.7%, 6.8%, 5.6%, 2.7% in the placebo, 100, 200 and 400 mg treatment groups respectively).

Small molecules including thalidomide congeners as well as MAPK inhibitors have a number of potential side effects including neurotoxicity and hepatotoxicity.

Careful adherence to a checklist (Table 3) can reduce adverse reactions to anti-TNF therapy, and most of these adverse events are likely to be an anti-TNF class effect rather than specific to infliximab.

References

Baert F, Norman M, Vermiere S et al 2003 Influence of immunogenicity on the long-term efficacy of infliximab in Crohn's disease. N Engl J Med 348:601–608

Braegger CP, Nicholls S, Murch SH et al 1992 Tumour necrosis factor alpha in stool as a marker of intestinal inflammation. Lancet 339:89–91

Braun J, Brandt J, Listing J et al 2003 Long-term efficacy and safety of infliximab in the treatment of ankylosing spondylitis. Arthritis Rheum 48:2224–2233
Cheifetz A, Smedley M, Martin S et al 2003 The incidence and management of infusion reactions to infliximab: a large center experience. Am J Gastroenterol 98:1315–1324
Colombel J-F, Loftus EV Jr, Tremaine WJ et al 2004 The safety profile of infliximab in patients with Crohn's disease: the Mayo clinic experience in 500 patients. Gastroenterology 126:19–31
Ehrenpreis ED, Kane SV, Cohen LB et al 1999 Thalidomide therapy for patients with refractory Crohn's disease: an open-label trial. Gastroenterology 117:1271–1277
Feagan BG, Sandborn WJ, Baker J et al 2000 A randomised, double-blind, placebo-controlled, multi-center trial of the engineered human antibody to TNF (CDP571) for steroid sparing and maintenance of remission in patients with steroid-dependent Crohn's disease. Gastroenterology 118:A655
Hanauer SB, Feagan BG, Lichtenstein GR et al 2002 Maintenance infliximab for Crohn's disease: the ACCENT I randomised trial. Lancet 359:1541–1549
Hommes D, van den Blink B, Plasse T et al 2002 Inhibition of stress-activated MAP kinases induces clinical improvement in moderate to severe Crohn's disease. Gastroenterology 122:7–14
Iyer S, Lahana R, Buelow R 2002 rational design and development of RDP58. Curr Pharm Des 8:2217–2229
Jewell DP, Lichtenstein GR, Gibson P, Innes A, Patel J 2003 CDP571, a humanised monoclonal antibody to TNF, is well tolerated in patients with Crohn's disease—a pooled analysis. Gut 52(suppl 6):A216
Lugering A, Schmidt M, Lugering N et al 2001 Infliximab induces apoptosis in monocytes from patients with chronic active Crohn's disease by using a caspase-dependent pathway. Gastroenterology 121:1145–1157
Murch SH, Lamkin VA, Savage MO et al 1991 Serum concentrations of tumour necrosis factor alpha in childhood chronic inflammatory bowel disease. Gut 32:913–917
Murch SH, Braegger CP, Walker-Smith JA, MacDonald TT 1993 Location of tumour necrosis factor alpha by immunohistochemistry in chronic inflammatory bowel disease. Gut 34:1705–1709
Papadakis KA, Targan SR 2000 Tumour necrosis factor: biology and therapeutic inhibitors. Gastroenterology 119:1148–1157
Parsi MA, Achkar J-P, Richardson S et al 2002 Prediction of response to infliximab in patients with Crohn's disease. Gastroenterology 123:707–713
Present DH, Rutgeerts P, Targan S et al 1999 Infliximab for the treatment of fistulas in patients with Crohn's disease. N Engl J Med 340:1398–1405
Probert CSJ, Hearing SD, Schreiber S et al 2003 Infliximab in moderately severe glucocorticoid resistant ulcerative colitis: a randomised controlled trial. Gut 52:998–1002
Ricart E, Panaccione R, Loftus EV, Tremaine W, Sandborn WJ 1999 Successful management of Crohn's disease of the ileoanal pouch with infliximab. Gastroenterology 117:429–432
Rutgeerts P, D'Haens G, Targan S et al 1999 Efficacy and safety of retreatment with anti-tumor necrosis factor antibody (infliximab) to maintain remission in Crohn's disease. Gastroenterology 117:761–769
Rutgeerts P, Lemmens L, Van Assche G et al 2001 Recombinant soluble p55 receptor induces remission, is non-immunogenic and well tolerated in active Crohn's disease: results of a randomised pilot trial. Gastroenterology 120:A452
Rutgeerts P, Feagan BG, Lichtenstein GR et al 2004 Comparison of scheduled and episodic treatment strategies of infliximab in Crohn's disease. Gastroenterology 126:402–403
Sandborn WJ, Feagan BG, Hanauer SB et al 2001a An engineered human antibody to TNF (CDP571) for active Crohn's disease: a randomised double-blind placebo-controlled trial. Gastroenterology 120:1330–1338

Sandborn WJ, Hanauer SB, Katz S et al 2001b Etanercept for active Crohn's disease: a randomised, double-blind, placebo-controlled trial. Gastroenterology 121:1088–1094
Sands B, van Deventer S, Bernstein C et al 2002 Long-term treatment of fistulizing Crohn's disease: response to infliximab in the ACCENT II trial through 54 weeks. Gastroenterology 122:A81
Scallon BJ, Moore MA, Trinh H et al 1995 Chimeric anti-TNF-alpha monoclonal antibody cA2 binds recombinant transmembrane TNF-alpha and activates immune effector functions. Cytokine 7:251–259
Schaible TF 2000 Long term safety of infliximab. Can J Gastroenterol 14(suppl C):29C–32C
Schreiber S, Rutgeerts P, Fedorak R, Khaliq-Kareemi M, Kamm MA, Patel J 2003a CDP870, a humanized anti-TNF antibody fragment, induces clinical response with remission in patients with active Crohn's disease. Gastroenterology 124(suppl 1):A61
Schreiber S, Winter T, Innes A, Patel J 2003b Safety of CDP870, a pegylated humanised anti-TNF antibody fragment, in Crohn's disease. Gut 52(suppl 7):A215
Targan SR, Hanauer SB, van Deventer SJH et al 1997 A short-term study of chimeric monoclonal antibody cA2 to tumour necrosis factor α for Crohn's disease. New Engl J Med 337:1029–1035
ten Hove T, van Montfrans C, Peppelenbosch MP, van Deventer SJ 2002 Infliximab treatment induces apoptosis of lamina propria T lymphocytes in Crohn's disease. Gut 50:206–211
Vasiliauskas EA, Kam LY, Abreu-Martin MT et al 1999 An open-label pilot study of low-dose thalidomide in chronically active, steroid dependent Crohn's disease. Gastroenterology 117:1278–1287
Vermeire S, Louis E, Rutgeerts P et al 2002 NOD2/CARD15 does not influence response to infliximab in Crohn's disease. Gastroenterology 123:106–111
Weinstein TA, Sciubba JJ, Levine J 1999 Thalidomide for the treatment of oral aphthous ulcers in Crohn's disease. J Pediatr Gastroenterol Nutr 28:214–216

DISCUSSION

Pavli: Are you aware of any health economic data comparing infliximab treatment with other therapies?

Ghosh: There are essentially two types of economic data. The first is the modelling which health economists use. The economic data, which are more relevant to us, include those of the type published by the Chicago group (Rubenstein et al 2002). These show that there is a significant reduction of hospitalization, surgery, endoscopy procedures and radiology following infliximab use, compared to the pre-infliximab period. In the UK, Derek Jewell has led the compilation of data for induction therapy and episodic use of infliximab. Again, this shows a reduction in hospitalization and possibly surgery. These translate to off-setting some of the cost of infliximab. We are still in the process of gathering robust economic data, especially for scheduled maintenance therapy. We haven't got any good economic data from the field (actual usage) of scheduled maintenance therapy apart from the subanalysis of the ACCENT I study, which as I showed reduces some hospitalization and surgery. Indirect costs have not been looked at at all, and we all know that this accounts for a major share of the total costs of managing this disease.

Jewell: The current data we have from 205 patients show that when direct costs are taken into account, the cost of infliximab per patient is GBP£8.60 more than otherwise. This makes it look like rather a cheap treatment, even without taking into account the indirect costs of time off work and the costs to society generally.

Rhodes: Are you talking about once-only treatment, or episodic, or maintenance?

Jewell: Of the 205 patients about a third of them had one infusion only and the majority had three or more. We looked at the direct costs in the six months prior to the first infusion compared with the six months following. It didn't address the question of whether patients were having episodic or maintenance therapy.

Schölmerich: There seems to be something like a health report system in the UK from the government. There is a study that has been published on the Internet (Clark et al 2003) using the ACCENT I data on the health economy of IBD and infliximab in particular. They came up with a figure of cost per QALY (quality adjusted to life year) of several tens of thousands of pounds. This is comparable to other relevant treatments. However, for the fistulae, the figure was much higher. They concluded that the fistula treatment was not economically reasonable while the general treatment was.

Jewell: I am not going to get into QALYs because the calculated QALYs varied from about £10 000 to £300 000 per year. When the group that produced the latter figure were asked to re-calculate their data they came down to about £29 000. It's not an exact science.

Peña: You didn't mention the topic of prediction. I know the data aren't there, but I'd like to hear what you think about this. All around the world people have now gathered enough data to show that there is a subgroup of patients who have late complications, and also autoimmune disease. Should the immunogenetics not just concentrate on finding genes that can predict response, but also ascertain whether there is a subgroup of patients who develop these complications?

Ghosh: With regard to the prediction of response, it remains difficult apart from the phenotypic prediction of response—those on immunosuppressives and perhaps the non-smokers appear to respond better. For some anti-TNF but not for all, perhaps elevated CRP (although we can debate the relevance of CRP) also better predicts response. Regarding the pharmacogenomic definition of response, we still haven't got a handle on this in a very robust way. This is an area that needs a lot more work. I am not aware of any direct connection of the pharmacogenomic approach to the development of future immune-mediated complications. Are you aware of any work in this area?

Peña: Not in this field. But in HIV, there was a recent editorial in *Nature* (Abbott 2003) commenting of the work of Roses who compared gene sequences in patients with and without toxic reactions to abacavir (Hetherington et al 2002). He identified three single nucleotide polymorphisms (SNPs) associated with immune-system genes that can be used to identify patients who shouldn't be given abacavir. This means that we now have the technology to take this kind of approach. I recently saw in New York a young man with severe heart failure due to autoimmune disease of the heart after several courses of infliximab. It is sad when this happens to a young person. Pharmacogenetics will help to prevent these severe reactions. The application of studies of SNPs might also have severe implications for insurance companies.

Jewell: We have had one patient who has a lupoid reaction to infliximab with a big rise in antibodies. He happened to be DR3 positive. It would be interesting to measure DR3 in all these patients who have or have not developed lupoid syndromes. At the end of the day, what have you demonstrated?

Rioux: It is a huge challenge. Just to be able to look at responder versus non-responder is one thing, but to be able to look at adverse reactions as a subgroup is hard. It is a good point, though: most of these studies don't have a DNA collection arm to them.

Ghosh: There is also the issue about how to define non-responders, i.e. as primary non-responders or non-responders later on. Lack of response can be overcome by dose manipulation in a proportion of patients.

Jewell: In general most of us believe that if you are non-responder to the first dose you stay non-responder.

Meddings: It strikes me that at least in Canada we are beginning to use this drug more and more. As we do this, we are increasingly using it on people with milder disease. I am slightly concerned that we are blasé about a 1–2% mortality rate for a group of patients for whom I would never have quoted a 1–2% mortality rate over many years with other therapies. Where are we going with this?

Ghosh: One of the reasons I presented these data was to raise that concern. In its defence, several of these patients who died in the Mayo series had significant and considerable co-morbidity. Nevertheless, these serious and fatal sepsis episodes mostly occurred within the first three infusions so they didn't appear to be dose related. One of the patients who died had an intra-abdominal abscess to begin with and should never have been given infliximab in the first place. Nevertheless there were completely unanticipated severe septic episodes in quite a significant number of patients.

Schölmerich: We have had a paper accepted for publication in *Gut* from Sweden, which looks at a population-based sample of patients treated with infliximab. The sample size is similar to that from the Mayo clinic study, although they didn't have such a select subgroup as in the Mayo. So this mortality figure of 1% may be real. One of the reviewers pointed out how similar the data are. If we start treating people top to bottom, that is if we start with the most aggressive treatment, even in the milder cases (which might be rational) we have to be aware of the problem that we might induce mortality.

Rhodes: With regards to mortality, we surveyed this in Crohn's disease pre-infliximab. Like others, we have found an increased mortality. The standardized mortality ratio works out at about 1.3. We shouldn't forget that Crohn's disease is a disease associated with mortality. This included about 33 deaths, half of which were related to the disease and half of which were unrelated. Of the half that were related to the disease there was a mixture of sepsis with and without operation and a small number with DVT and pulmonary embolism. You will have seen a nice paper in *Gastroenterology* about three months ago (Aberra et al 2003) showing that steroids were associated with an increased rate of postoperative sepsis in CD patients. We have recently completed a study showing that steroids are associated with increased risk of spontaneous sepsis not associated with surgery. Putting these two things together, the 1% drug-associated mortality would need to be taken into the context of the underlying mortality of Crohn's and the timescale to accrue that 1%. But I would also question whether some of that might be due to concomitant use of steroids: it is common practice to give these first and if they don't work then to use infliximab. So you have already given a sepsis-inducing trigger with a steroid before adding infliximab. I would suggest that we should be looking hard at using infliximab without steroids. I can't see a good reason for using steroids in a patient who has had a relapse on azathioprine or methotrexate.

Sartor: I'd like to explore the immunogenicity. I am struggling with the decision to treat once versus three times. The argument for treating with three infusions for induction is (1) increased healing and (2) the decreased antibody response. In one study there was a comment in the text (but not in a figure) that from the single luminal infusion, the antibody response was one-third that of the three infusions. This would dispute the idea that giving three infusions actually decreased antibody responses. In the ACCENT I follow-up 46% of patients could not be commented on because they had circulating infliximab. This would probably be only in the scheduled groups. Yet in the follow-up study there are good data that there is no difference in infusion reactions that were highly correlated with circulating antibody levels. Loss of response was similar in those groups. As you mention, there was a follow-up of between 30 and 50 weeks. My conclusion is that the antibody rate is probably being understated in the group that got the continued infusions versus the episodic group, and that there wasn't a big difference in antibody response. The problem is that in the methods it was stated that they measured antibodies at 72 weeks, but that was not reported.

Ghosh: My problem with the one dose versus three dose induction is that immunogenicity doesn't play a major role in that at all. It is increased serum levels that are achieved by infusing them at 0, 2 and 6 weeks that leads to a higher efficacy rate. I find it difficult to accept some of the contention that this is due to reduced immunogenicity. The difficulty with reconciling the ACCENT I data with the kind of data that Paul Rutgeerts has produced on immunogenicity with ATI is severalfold. Apart from the manner in which you use it, I am not sure to what extent the Centocor antibody assay is the same as the Prometheus assay which Paul Rutgeerts used to measured the antibody response. These two assays don't detect the same epitopes. I have a feeling that unless we get more quantitative data about the antibodies, it will be difficult to interpret the response, particularly given that the antibody response can only be evaluated in a small proportion of the patients. In the rest of them the presence of infliximab interferes with it. In most of the Centocor antibody assays the lack of quantitative determination, coupled with the fact that a lot of them are described as indeterminate because of the presence of infliximab, makes it very difficult to assess the role of immunogenicity in the response. The way we aim to overcome this immunogenicity, by pouring in more infliximab, either in quantity or frequency, seems to be slightly medieval. There must be other ways of inducing tolerance than just pouring in increasing quantities of antibody. The trouble with all the Centocor data has been this in-house assay that is slightly uncharacterized and is not quantitative. A lot of the patients don't have the test anyway.

Sartor: When you have a patient which you need to induce with luminal disease, do you use one or three doses?

Ghosh: I don't use either; I use one dose at 0 and another dose at 4. This is a compromise. The whole dosing schedule is complicated by the fact that in rheumatoid arthritis 3 mg/kg doses are used, and in Crohn's 5–10 mg/kg is used. In ankylosing spondilitis a 5 mg dose is used but at 6 weeks rather than 8 weeks. If we look at the pharmacokinetics of the antibody use at 8 weeks is likely to mimic an episodic rather than a maintenance use. To claim that it is scheduled maintenance is an overstatement of the immunological pK_a of the drug.

Pavli: In rheumatoid arthritis we are giving concomitant immunosuppressive agents in the form of methotrexate, so we are comparing apples with oranges in that situation.

Sartor: The true message is that concomitant immunosuppressives must be given, to decrease immunogenicity and, I hope, to increase efficacy. The problem is that all of these studies are exclusively looking at failures, and are not changing the dose to optimize. We, as a community, need to do a study in which we prospectively look at single or multiple infusions with and without concomitant immunosuppression.

Pavli: We are trying to develop such a study in Australia.

Rhodes: I was very interested by the data on hospital admissions and surgical rates in intermittent versus maintenance therapy. Has this been analysed for the patients who are also taking azathioprine or methotrexate? In other words, if you are on concomitant immunosuppressives, what is the best course? Should we be treating these episodically or by maintenance?

Ghosh: It didn't come out as a significant factor in that analysis, but this doesn't exclude the possibility.

Jewell: In our 205 patients, 75% were taking concomitant azathioprine and a few were on methotrexate.

Rhodes: But for the patients taking these two drugs, what is the best course? Episodic or maintenance treatment?

Jewell: That is an unsolved issue. I want to raise another point, concerning mechanisms of action. The initial story was that CDP571/870 did not induce apoptosis. My understanding now is that it does. Sander Van Deventer, do you have any data comparing CDP571/870 with infliximab in terms of apoptosis, and would you like to go on and say where this apoptosis takes place?

Van Deventer: For infliximab we know quite well. In patients it induces apoptosis of $CD3^+$ T cells only in the mucosa and not in the circulation. We have done extensive analysis taking samples every 2 h for 24 h and there is nothing there. *Ex vivo* we think this is related to binding of the antibody to T cells that express membrane-expressed TNF. This is a specific subpopulation of Th1 cells. These are in the mucosa. We know this occurs *in vivo* because we have done scans in mice and humans with an anti-TNF antibody. It is a 100% hit. A major issue is that Enbrel (Etanercept) is not very effective. If you talk to the people at Amgen they say that maybe Enbrel is effective at a higher dose. We are therefore trying high dose Enbrel and scanned subjects to see whether there is apoptosis. Most of the pharmaceutical companies chicken out. One other issue that confounds this discussion is the complement-mediated story. As it turns out, the apoptosis *ex vivo* is completely independent of complement, so this is not the mechanism.

References

Abbott A 2003 With your genes? Take one of these, three times a day. Nature 425:760–762

Aberra FN, Lewis JD, Hass D, Rombeau JL, Osborne B, Lichtenstein GR 2003 Corticosteroids and immunomodulators: postoperative infectious complication risk in inflammatory bowel disease patients. Gastroenterology 125:320–327

Clark W, Raftery J, Song F et al 2003 Systematic review and economic evaluation of the effectiveness of infliximab for the treatment of Crohn's disease. Health Technol Assess 7:1–67

Hetherington S, Hughes AR, Mosteller M et al 2002 Genetic variations in HLA-B region and hypersensitivity reactions to abacavir. Lancet 359:1121–1122

Rubenstein JH, Chong RY, Cohen RD 2002 Infliximab decreases resource use among patients with Crohn's disease. J Clin Gastroenterol 35:151–156

Final discussion

Jewell: I thought in this final discussion it might be best to begin by running through one or two themes that have been raised over the last few days. We have come a long way in the 100 years since the meeting at the Royal Society of Medicine in 1905, where they debated hotly whether the aetiology of ulcerative colitis (UC) was due to something in the diet, a reaction to bacteria, or was psychological. We have had a fascinating discussion about the role of diet and how it can alter the flora, and affect epithelial function. There are clearly a lot of unanswered questions about this, and the intimation we had in the discussion this morning about how events in the epithelium can affect bacteria is fascinating. We touched on the mechanism by which pathogens will interact with a number of specific pathways in the epithelium. What we didn't really get into is whether the good bacteria — the lactobacilli and bifidobacteria and so on — have any direct interaction with epithelial cell function. This might be quite an interesting area to explore and we may have some more discussion on this. We haven't really discussed the psychological aspects, but Salvador Peña briefly raised the issue of stress earlier on. We have ignored the whole of the neuroendocrine control of the immune response through lack of time. It is a fascinating area. We are beginning to get some good science on how stressful events could potentially affect gut immune function, permeability and so on. The big difference between 100 years ago and now is that we are now very much more informed about host factors. We have heard about some elegant studies on the barrier function of the epithelium during this meeting, about how cells join together to form a nice barrier. Despite the fact that Jon Rhodes has been with us we haven't said a word about mucus. I am not convinced that the abnormal permeability that is being discovered in first degree relatives of our patients is an environmental response. I would like to think that this reflects a genetic factor that is operating within these families. Underlying this immune response, epithelial function and interaction with bacteria, is the host genetic response. We are going to get a lot of messages coming through. NOD2 and Paneth cell function is a fascinating story. How relevant it is will be determined in time when the observations have been repeated. We are about to present some evidence of interleukin (IL)10 receptor polymorphisms in patients with Crohn's disease and UC. These polymorphisms are amino acid changing and are therefore likely to be functional. It is quite possible that however much IL10 one might pour into a patient, if they have the wrong sort of receptors they are not going to respond. This is another area that is going to open up into a fascinating story if it is confirmed. This brings us back to the pharmacogenetic area. How might this go forward? At the moment it is the big promise of therapy. We have covered a lot of these points. We have opened up a number of proposed collaborations between people in this room, which is an exciting development. We have pointed to the future in a number of different areas, from genomics through to gene expression studies using microarrays and proteomics. We have even heard about kinome profiling, which is fascinating, and we have heard about new ways of studying immune regulation, particularly with dendritic cells (DCs) and T_{reg} cells. It has been a fascinating few days from which I have learned a lot. Are there any vital points arising from this?

Roediger: I want to raise the question of the barrier. In Crohn's disease and UC we have breakdown of the barrier. We have certain concepts that activated immune cells are crucial to this. Beyond this activation of the immune cells, how is the barrier broken down? When I ask my colleagues they mostly say it is cytokines. It is a question I'd like to put to the immunologists, microbiologists and biochemists.

Jewell: Perhaps we should start with Chuck Parkos, because he produced some nice data showing how interferon (IFN)γ and tumour necrosis factor (TNF) interact to disrupt barrier function.

Parkos: Cytokines are one way that the barrier gets disrupted. I don't know whether we are looking at the chicken or the egg here. John Rioux's genetic analyses would strongly indicate that it is going to be multifactorial. Immune cell dysregulation leads to cytokine production, and there is abnormal response of the epithelium to not only cytokines but also a range of stressors. How this epithelium responds might cause a disrupted barrier that then allows the floodgate to open and this vicious cycle to perpetuate itself. We are just barely scratching the surface on the endocytosis story, for example. The nature of the pathways that regulate turnover of proteins that regulate permeability is a huge subject that little is known about. It was mentioned earlier that even Toll receptors are present in vesicles in epithelial cells, and, whether bacterial products can regulate epithelial cell function at that level needs to be explored. To me this is frustrating and exciting at the same time. It reinforces the idea that many different inputs might give the phenotype of inflammatory bowel disease (IBD).

Jewell: Let's go on exploring this a little. A few years back we thought it would be fascinating to look at some of the polymorphisms in these proteins in the tight junction and adherens junction. We were a little distressed to find that at the time most of these molecules weren't thought to be very polymorphic. Then people started discovering molecules such as the claudins at such a rate that the genotyping couldn't keep up. Do you think there could be a genetic component controlling intercellular adhesion, or do you think that any functional polymorphism would have such a lethal effect that the individual wouldn't actually survive?

Parkos: There are now mutations that have been described in some of the claudin family members that are distinct entities. For example, there is congenital hypomagnesaemia, and this results from a mutation in claudin 11. This protein is expressed exclusively in the kidney and the mutation results in magnesium loss in the urine. This is an example of a junction protein that, when mutated, is not lethal but has a clinical syndrome. There is another claudin mutation that causes a congenital form of deafness and has to do with regulation of permeability in the inner ear. Jim Anderson showed very nicely the variability of claudin expression in the gut. I don't know whether mutation or alteration of one of these might cause some subtle defect in the regulation of permeability that could explain at least some of these patients. It is possible therefore that you could have a defect in one of the structural components of the tight junctions with functional consequences. The problem is that we haven't even characterized all the components of the tight junction. Take cocksackie adenovirus receptor as an example. If you do an expressed sequence tag (EST) search there is a protein called cocksackie adenovirus-like receptor. A paper came out recently showing that this protein clearly localizes to the tight junctions, and no one has any idea what it does, except that it fits in the immunoglobulin superfamily category (Raschperger et al 2004). It is possible that these proteins bind scaffolding proteins such as ZO1 and AF6, and yet they may not form the seal. Tsukida proposed that these IgSFs are recruited to the intercellular junctions in a similar fashion to E-cadherin along with the scaffolding proteins to form the meshwork for tight junctions. Again, this increases the complexity: any one aspect of this pathway might lead to aberrant regulation of junction permeability. The JAM(A) story is like that of E-cadherin in that if you disrupt the barrier and add functionally inhibitory antibodies to these proteins, you can significantly delay the recovery of the functional barrier. However, if you look at the structural elements, they seem to come back yet the barrier remains disrupted. This tells me that we don't have a good handle on what all of the players are. I would like to use a kinase assay to do these experiments and see which kinases are activated. I suspect that we will see specific ones that are important in regulating the assembly of intracellular junctions.

Jewell: Liz Furrie, do you think that the 'good' bacteria do something directly to the epithelium that is beneficial to the host?

Furrie: Yes. There is an increasing body of evidence suggesting that these bacteria directly interact with the epithelial barrier. Also, there are tantalizing data concerning gap junctions. There are data showing that certain commensals might help to strengthen gap junctions, and you could imagine that someone will have a propensity towards IBD if they have a slight mutation in one of the gap junction proteins which makes them leaky, but not pathologically leaky, and then you have a bacterium which gets into the wrong niche. This could produce something that allows them to disrupt another one of the interactions. You could imagine that this would initiate breaking down of the system. The 'good' bacteria could come in and replace the inappropriate colonizer in this niche and help build up gap junction integrity again.

Parkos: I may have seen a paper or abstract (I don't remember which one) talking about increased transepithelial resistance with some of these beneficial bacteria.

Meddings: There are several examples of this. There could easily be a genetic defect in tight junctions. It is a complicated field. The thing that has amazed me is that over the last five or six years what has emerged is all of these interactions between the organisms that live on the surface of the epithelial cells. They can produce toxins that attack claudins and tight junction proteins, they can directly adhere to epithelial cells, they can insert bacterial proteins that can be their own receptor, and they can usurp intracellular processes that change metabolism and affect tight junctions. Beth McCormick has also shown that there are non-pathogenic bacteria that can inhibit ubiquitination of I-κB and therefore inhibit this process. The interactions are amazingly complex. It might be a change in flora that can talk to bugs, or a change in host genetic make-up that produces another population of bugs. There is an interplay that we haven't really recognized until now.

Furrie: The issue of quorum sensing in complex microbial communities is difficult to study but is essential to our understanding of how to control the growth of the community rather than an individual organism. This is why mono-association is good at showing trends but is not what actually really happens. People are going to have to try to bi- and tri- and multi-associate in a sequential manner.

Sartor: I agree. For all of our mono-association work, although it is an elegant way to dissect out components, none of it recapitulates the more aggressive disease in a complex environment. I want to put out a plea that we not only think about bacteria as species and strains, but we also think about virulence factors within bacteria. You can grow *Escherichia coli* but this is the tip of the iceberg: it may have adherence genes, it may secrete toxins and it might put out type 3 secretion systems that paralyse ubiquitination and proteasome activity. It can have all kinds of effects. We need gene arrays of bacteria.

Furrie: If you spoke to microbiologists 20 years ago they would have said that *E. coli* was a normal commensal organism.

Kelleher: If you look at where there is mono-association, in the stomach when there is *Helicobacter* present, the amount of influence this has on the epithelial cells is enormous. We have already heard about type 4 secretion and the influence on zone 1, but there are changes on epithelial permeability that you can induce with *H. pylori* that are not cagA-dependent at all. We know that it induces NF-κB and we now have evidence that thioredoxin is secreted by *H. pylori* and it is a potent NF-κB antagonist. It has two ways of operating with respect to the epithelium. If you then start to look at complex populations of bacteria it is going to be incredibly difficult to sift out the true mechanisms in IBD. You might be looking at multiple additions of very small effects on epithelial barrier function. This brings me to one more effect we didn't really discuss, which is diet. One of the critical issues in diet is that we can alter fatty acid membrane composition by altering the fatty acid composition of the diet. Some of the kinases, such as protein kinase C, are modulated by fatty acids. We don't have a handle on how fatty acids can affect tight junction function except in a few isolated cases.

Rhodes: Can I respond to your gentle goad about mucus and give a brief update on it? It is a very big subject. The most consistent abnormality that has been found is increased expression of mucosal oncofetal carbohydrate antigens which are short disaccharides that are not normally expressed. They become exposed in IBD and cancer. They occur not only on the mucins but also on cell surface glycoproteins. We have shown that the Thomsen–Friedenreich (TF) antigen and sialyl-Tn, which are the two best studied carbohydrate antigens, are predominantly expressed on high molecular weight splice variants of CD44 on the cell membrane (Singh et al 2001). This in itself is quite interesting because it is well established and these splice variants are up-regulated in IBD and in colon cancer, driven by pro-inflammatory cytokines. It is also particularly interesting because the high molecular weight splice variants of CD44 are now known to associate with c-Met, which is the HGF receptor. When they associate this allows c-Met to activate the MAP kinase pathway, which it can't do when they are not associated. This raises the possibility that ligands that interact with that carbohydrate structure on CD44 might have an impact on that pathway. Indeed, we have found that peanut lectin does phosphorylate c-Met and then activates that pathway. It seems conceivable that bacteria might be capable of doing the same thing. We have only so far managed to find one *E. coli* isolate with lectins specific for TF antigens, so this seems to be a very rare phenomenon, but we have found one cancer-related *E. coli* isolate which does that. Another mechanism for the altered glycosylation seems to be altered Golgi pH. This has been reported in cancer and needs looking at in IBD. This is probably driven by pro-inflammatory cytokines as well. You lose the normal acidification of the Golgi. Whether or not any of this is related to any genetic abnormality in IBD is unproven. Dan Podolsky did an interesting study with Curt Tysk where they did an ion exchange chromatographic separation and showed a subclass defect in the unaffected identical twins of IBD patients. Curt Tysk has very kindly sent us some sections from those biopsies and we have shown that it is possible to identify glycosylation abnormalities in the Golgi region of the surface epithelium of those unaffected identical twins. Interestingly, this is just confined to the surface epithelium and is not present lower down the crypts. To my mind this makes it much more likely that this is an acquired phenomenon rather than genetic. The question then is what is causing it, because there aren't any neutrophils to see. I think this would have to be some extremely subtle early 'inflammation' that is not driven by the neutrophils. It raises the possibility that Brandtzaeg's hypothesis for UC might be plausible: that it is some sort of luminal-driven immune-mediated event which is able to trigger NF-κB activation in the surface epithelium before there is encroachment with neutrophils. This story also has relevance to bacterial adherence. We have shown that the *E. coli* associated with the mucosa in IBD and colon cancer are more likely to express haemagglutinins and are more likely to be able to adhere to intestinal epithelial cell lines.

Sartor: You might be interested in some of our data. Germ-free animals have decreased MUC2 expression and a different glycosylation and sulphation pattern than animals that are colonized with normal bacteria. Intriguingly, the IL10 knockout mouse, even in the germ-free state, has decreased MUC2 expression. One of the features that Fiona Powrie raised in her presentation was the decreased number of goblet cells. There is probably not a total annihilation of goblet cells, but they are not packed with mucus — they are actually secreting it. Certainly, the inflammation state and bacterial state will strongly influence expression, sulfation and glycosylation of mucus.

Rhodes: That's true. One thing that was sad to us mucus aficionados was the lack of colitis in the MUC2 knockout, which develops colon cancer. I am not sure whether that knockout has been looked at yet with pathogens, to see if it is more susceptible.

Powrie: Has it been infected with *Helicobacter hepaticus*?

Rhodes: I don't know.

Roediger: *E. coli* has been mentioned with respect to colitis. Is *E. coli* more plentiful on the mucosa in colitis? It is such a versatile genetic organism.

Furrie: Numbers do go up on the epithelium but not dramatically. Just saying they have gone up a bit doesn't really mean anything, though, because you don't know whether they have gone up because of the inflammation or the disease aetiology.

Roediger: And you don't see it in the early stages of the disease.

Furrie: How do you know that you are looking at the early stages of disease?

Sartor: Swidsinski showed that *Enterobacteria eastiae* was one of the bacterial groups that increased in his mucosal-associated paper in *Gastroenterology* last year (Swidsinski et al 2002). We have shown that if you colonize IL10-knockout mice on a 129 background, and also 129 germ free animals with exactly the same flora, there is a drift in the presence and absence of inflammation such that *E. coli*, *Klebsiella* and *Clostridium ramosum* expand, and two other broad clostridial categories decrease. Whether this is a secondary phenomenon or not I don't know. Certainly with a murine *E. coli* mono-association, the *E. coli* knockout animals get colitis.

Schölmerich: I would like to turn the whole discussion around. We are always discussing why the epithelium gets leaky. But surely the important question is why do the 99.9% of people who don't get Crohn's and colitis tolerate their flora? This is an interesting issue. We have only one immune system at the time we learn to tolerate the flora, which is the innate immune system; the adaptive system is still developing. The innate immune system is very complex. Don't you think that we have this broad area of pattern recognition receptors, such as the Tolls, which may be there to recognize the good ones and not the bad ones? They might be signalling to the good bacteria that they can stay. The problem to me seems to be here. The adaptive immune system must drive the secondary problem of general inflammation because it is not there when we learn the most important thing.

Furrie: The expression of these pattern recognition receptors in a maturing gut is not clear. No one has looked at the dynamics or expression with development. This is a huge black box.

Moore: You can expand the issues that we are discussing here to other immune-mediated diseases since the routes through the gut epithelium and the host recognition processes will affect the way many antigens gain access. The issues you are working on are not just important for IBD, but are also important for many other diseases.

Ghosh: On this general theme, we are working on the concept of chronic immune-mediated diseases. But to move back to the gut, in comparative immunology Coeliac disease is a Th1-driven crypt hyperplasia villus atrophy model with a leaky gut similar to Crohn's disease. Despite occasional overlap the phenotypic manifestations of Coeliac disease and Crohn's disease are so different. There is a lot of merit in trying to understand not just IBD, but the private components of these individual diseases and the public components of a chronic inflammatory process. We can disrupt the barrier in many ways, but nevertheless we do not always end up getting IBD.

Jewell: It is time to bring this meeting to a close. The subtitle of this meeting was 'crossroads of microbes, epithelium and immune systems'. This has led to a tremendous amount of cross-talk between us that represents a very exciting interactive discussion. From a patient's point of view the exciting thing over the last few years has been that this focusing on mechanisms of inflammation and disease progress has actually produced new therapies. This momentum is accelerating rather than decelerating. We have also had discussion that has raised new avenues for research. We have only just touched the surface and the field is developing so rapidly that we might be holding this meeting again in a few years time. Thank you all for your participation in this meeting.

References

Raschperger E, Engstrom U, Pettersson RF, Fuxe J 2004 CLMP, a novel member of the CTX family and a new component of epithelial tight junctions. J Biol Chem 279:796–804

Singh R, Campbell BJ, Yu LG et al 2001 Cell surface-expressed Thomsen-Friedenreich antigen in colon cancer is predominantly carried on high molecular weight splice variants of CD44. Glycobiology 11:587–592

Swidsinski A, Ladhoff A, Pernthaler A et al 2002 Mucosal flora in inflammatory bowel disease. Gastroenterology 122:44–54

Index of contributors

Non-participating co-authors are indicated by asterisks. Entries in bold indicate papers; other entries refer to discussion contributions.

Subject index

Page numbers in *italic* indicate tables.

J

L

M

Printed and bound by CPI Group (UK) Ltd, Croydon, CR0 4YY

16/09/2026

14963379-0002